Rory McCloy (Ed.)

Quality Control in Endoscopy

Report of an International Forum held
in May 1991

Contributors
James Whitwam · Eran Geller · Emmet Keeffe
David Fleischer · Alan Maynard · Nicola Davies
David Poswillo CBE

Springer-Verlag
Berlin Heidelberg New York
London Paris Tokyo
Hong Kong Barcelona
Budapest

Dr. Rory McCloy BSc MD FRGS
University Department of Surgery
Manchester Royal Infirmary
Manchester M13 9WL
United Kingdom

Editorial Co-ordinator
Anne Pringle Davies BSc

This International Forum was funded by an educational grant from
F Hoffmann-La Roche Ltd, Basle, Switzerland

ISBN-13:978-3-642-77140-8   e-ISBN-13:978-3-642-77138-5
DOI: 10.1007/978-3-642-77138-5

The use of general descriptive names, registered names, trademarks, etc. in this publication does not imply, even in the absence of a specific statement, that such names are
exempt from the relevant protective laws and regulations and therefore free for general
use.

Product liability: The publishers cannot guarantee the accuracy of any information
about dosage and application contained in this book. In every individual case the user
must check such information by consulting the relevant literature.

Type setting: Cicero Lasersatz, 8900 Augsburg

25/3140/543210 – Printed on acid-free paper

Contents

Section 4 – Medico-legal Aspects

Forum President

Dr. Rory McCloy BSc MD FRCS
Senior Lecturer and Honorary Consultant
University Department of Surgery
Manchester Royal Infirmary
Manchester
United Kingdom

Secretary of the Endoscopy Section, British Society of Gastroenterology (BSG)
Secretary of BSG Working Party on Sedation and Monitoring for Endoscopy
Chairman of BSG Working Party on Informed Consent for Endoscopy
Project Leader, Audits of appropriateness and adverse outcomes of upper gastrointestinal endoscopy, a Joint Project of the Royal College of Physicians of London, BSG, Association of Surgeons of Great Britain and Ireland, College of Anaesthetists, Society of Thoracic Surgeons of Great Britain and Ireland.

Editor's Preface

Surprisingly, little attention has been paid to the broader issues of patient care related to endoscopy. Usually, meetings and reviews concerned with the process of endoscopy have confined themselves to specific and traditional aspects, such as endoscopic techniques and applications. These areas are well researched, widely practised and fully documented elsewhere.

An international Forum, a marketplace with the emphasis on discussion, was conceived to embrace aspects of endoscopy which are crucial to safe practice and consistent with quality of care for our patients.

Unlike traditional symposia where pundits lecture to relatively passive audiences, the Forum created an open environment where information and opinions were shared. All the delegates were active participants and selected from 17 countries for their knowledge and expertise in endoscopy and related disciplines.

It may seem unusual at a meeting on Quality Control in Endoscopy that only three members of the faculty were practising endoscopists. Since we were discussing four main areas – drugs for sedation, procedural safety, resource management and medico-legal aspects, specialists in each of these fields were invited to contribute their particular expertise and apply it to the topic under discussion, endoscopy.

The innovative format of the Forum, which met in May 1991, deserves an explanation. The meeting was divided into three parts. The opening session consisted of keynote lectures on each of four main topics.

The Forum then divided into simultaneous workshops dealing with each of these areas. In the Workshops dealing with Drugs for Sedation and Procedural Safety, the participants had each been pre-assigned a topic. Their preparation prior to the event enabled them to act as lead-off discussants on these topics within the Workshops. Forty eight participants contributed to fourteen hours of in-depth discussion within the Workshops.

Participants in the Workshop on Resource Management were sent a questionnaire dealing with staff, drugs, consumables, equipment and overheads – the main cost components of endoscopy. The results from these highlighted many of the problems of trying to apply health economics to a specific area in which data are lacking.

For the Medico-legal Workshop, two legal cases were devised. Delegates in this Workshop were asked to prepare themselves to act as expert medical witnesses in defence or prosecution of the endoscopist on trial. Discussions arising from this covered the principal issues of negligence and informed consent.

Each of the four Workshop Directors then presented a summary of their Workshop discussions, conclusions and recommendations at a final plenary session.

This publication presents the four keynote lectures and their complementary Workshop Reports, combined to form a unique overview of quality control in endoscopy.

RORY McCLOY
University of Manchester
September 1991

Section 1: Drugs for Sedation

Keynote Lecturer

Professor JAMES WHITWAM MB ChB PhD FRCP FFARCS
Director of Anaesthetics
Royal Postgraduate Medical School
Hammersmith Hospital
London
United Kingdom

Workshop Director

Professor ERAN GELLER MSc MD
Director Department of Anesthesiology and Critical Care Medicine
Tel Aviv Medical Center

and

Associate Professor of Anesthesiology
Chairman, Division of Anesthesiology
Sackler Faculty of Medicine
Tel Aviv University
Tel Aviv
Israel

Workshop Participants

Dr. J. R. ARMEGOL MIRO, Barcelona, Spain
Professor R. ARNOLD, Marburg, Gemany
Dr. J. F. W. BARTELSMAN, Amsterdam, The Netherlands
Dr. K. BAUCH, Chemnitz, Germany
Dr. F. EIGENMANN, Berne, Switzerland
Dr. A. GEVERS, Leuven, Belgium
Dr. B. HOFFSTAD, Oslo, Norway
Dr. E. K. JANATUINEN, Kuopio, Finland
Mr. P. N. J. LANGENDIJK, Amsterdam, The Netherlands
Dr. A. MALDONADO, Lisbon, Portugal
Dr. R. PEARSON, Manchester, United Kingdom

Dr. F. SERVIN, Paris, France
Dr. M. VAN BLANKENSTEIN, Rotterdam, The Netherlands
Dr. S. D. J. VAN DER WERFF, The Hague, The Netherlands
Dr. F. VICARI, Nancy, France

CHAPTER 1

Drugs for Sedation

Professor JAMES WHITWAM

Introduction

Modern endoscopic techniques are being practised by a progressively wider spectrum of specialists for a greater range of both diagnostic and increasingly complex therapeutic procedures, particularly in the fields of gastroenterology, urology [1], gynaecology and cardiology, by orthopaedic and vascular surgeons, and interventional radiologists.

Such procedures are at best unpleasant and can be extremely disturbing for many patients, causing anxiety amounting to fear and feelings of panic, particularly when the airway is compromised as may occur during endoscopy. Fear more than pain is a major stimulus to increased sympathetic activity, causing hypertension, arrhythmias, an increase in the myocardial oxygen demand to supply ratio with myocardial ischaemia, the risk of myocardial infarction and cardiac arrest which will be compounded by hypoxaemia.

Apart from anxiolysis, amnesia is also a desirable state in such patients ensuring that they have no fear of returning for further investigation or treatment. Arguably in this context a high degree of amnesia is equivalent to anaesthesia, with the advantage of maintaining verbal communication and avoiding the anaesthetic state with its complications, responsibilities and medico-legal implications.

Sedo-analgesia

Whenever possible, pain at the site of intervention, and hence nociceptive reflexes, should be obtunded by using a local anaesthetic drug. However, even under these circumstances, complete management of the patient includes sedation, not only to allay anxiety but also if the procedure is prolonged or the position uncomfortable. A combination of local analgesia and sedation is termed sedo-analgesia [1].

When pain occurs and local anaesthesia is not feasible, for example in colonoscopy and some gynaecological procedures, such as oöcyte retrieval, it may become necessary to consider the administration of a centrally acting analgesic drug, e.g. fentanyl or alfentanil, which has a synergistic interaction with the sedative drug and may cause deep sedation or induce anaesthesia with loss of verbal communication and the airway, and also increase respiratory and cardiovascular depression. At any subsequent enquiry such an event implies that the practitioner and assistant are competent in the conduct of anaesthesia and resuscitation to standards required by specialists in anaesthesia [2]. Polypharmacy should by avoided except by suitably trained practitioners [2].

Objectives of sedation

1. Behavioural. Anxiolysis, reduction of attention, amnesia, maintenance of verbal communication and cooperation by the patient.
2. Physiological. Retention of a brisk glabellar eye blink response, reduction of sympathetic activity and possibly also muscle tone.
3. Monitoring. Arterial oxygen saturation (SaO_2) greater than 90% or within 5% of control values.

Conduct of sedation for gastroscopy

External stimuli can reduce the effectiveness of sedative drugs. The environment should be orderly and quiet with carefully planned illumination such that the total candela in the patients' view are reduced to only a necessary minimum. After initial work up and stabilisation of sedation, repeated cuff blood pressure measurements are best avoided, and other monitoring equipment should be muted apart from alarms.

However, in view of the currently reported mortality rate, there should be *pre-requisites before sedation for gastroscopy is commenced.*
a) Availability of continuous monitoring with both a pulse oximeter and ECG.
b) Both the practitioner and assistant should have recently certified evidence of training in cardiopulmonary resuscitation.
c) Resuscitation equipment, i.e. oxygen delivery and ventilation systems, suction, defibrillator and drugs required for resuscitation should be immediately to hand.
d) There should be the possibility of rapid transfer to full critical care facilities (e.g. intensive care, operating theatres, transfusion).
e) An intravenous cannula should be introduced to allow continuous venous access.

Drugs available for sedation

All drugs used for the induction of anaesthesia by the intravenous route can be administered in smaller doses to provide sedation. Inhalational anaesthetic agents with analgesic properties, e.g. nitrous oxide, have been used widely in obstetrics and dentistry, but their use is limited and in decline and they will not be discussed here. Anaesthesia is induced intravenously by four principal drug types and a variety of combinations of these.

1. *Classical intravenous induction agents,* effective within one arm-brain circulation time, e.g.
   a) *barbiturates* – thiopentone, methohexitone,
   b) *imidazole derivative* – etomidate,
   c) *phenol derivative* – propofol.
2. *Opioids* – morphine, fentanyl, alfentanil.
3. *Phencyclidine derivative* – ketamine.
4. *Benzodiazepines* – midazolam, diazepam.

Other drugs such as droperidol are sometimes used as adjuncts to the other groups of drugs.

With the exception of the benzodiazepines and the opioids, the therapeutic window available for sedation, before the induction of anaesthesia, can be extremely narrow. For example, propofol is a highly potent and extremely short-acting anaesthetic drug which is playing a rapidly increasing role in anaesthetic practice, including intensive care, particularly when administered by infusion [3–5]. Sedation, light and deep anaesthesia occur at plasma levels of 2, 3 and 6 $\mu g.ml^{-1}$ respectively. Thus it is one thing to achieve and maintain anaesthesia but, in view of the individual variation in response of patients to the drug, it is much more difficult to quickly establish and maintain a constant level of adequate sedation, e.g. within 5 min, with no risk of trespassing into anaesthesia either at the start or at a later stage. Propofol can cause apnoea and severe cardiorespiratory depression in anaesthetised patients although intraoperative patient controlled sedation with propofol by infusion has been described [6]. However it remains an anaesthetist's or intensivist's drug and at present its use requires at the very least monitored anaesthetic care (MAC), i.e. the presence of an anaesthetist. These remarks are particularly apposite to gastroscopy with the need for a rapid turnover of patients and where the presence of anaesthetists on the one hand and personnel trained, practised and certified in anaesthesia and resuscitation on the other, are the exception rather than the rule.

Opioids may be used alone to cause analgesia and sedation, but they can cause severe respiratory depression, bradycardia and muscle rigidity. They always introduce an incidence of nausea and vomiting. However their current safe application in patient controlled analgesia (PCA) implies that they can be used providing the doses and conditions of administration are carefully controlled.

Ketamine produces a state called "dissociative anaesthesia" and has a relatively long duration of action. Its side effects include hypertension, tachycardia, a rise in intracranial pressure and severe, unacceptable, dysphoric effects [7]. Although these can be ameliorated by benzodiazepines such side effects preclude its use in routine clinical practice. It is an invaluable drug for the induction of anaesthesia with only minimal cardiovascular and respiratory depression in difficult or emergency situations, e.g. in radio-diagnostic and therapy departments and in the trauma services. It is also widely used by many anaesthetists, often in combination with midazolam, for the induction of anaesthesia for major procedures, particularly in critically ill patients [8, 9].

The benzodiazepines (BZDs), because of their wide margin of safety and ease of administration, are used not only by anaesthetists but also by a wide spectrum of practitioners, both medical and dental, with little or no formal training in anaesthesia. The reason is that given good patient selection, an appreciation of the sensitivity of the elderly to sedative drugs, knowledge of the recommended doses in normal subjects, and above all an element of intuitive good sense, it is possible to practise safe sedation with little risk of inducing anaesthesia.

The development and introduction of midazolam provided for the first time a short-acting water soluble benzodiazepine which, by providing in addition a high level of amnesia and virtually no enterohepatic circulation in normal subjects, represents a major advance in sedation therapy. The subsequent development and introduction of flumazenil now allows the termination of benzodiazepine-induced sedation at will, which has major implications not only for safety but also potentially for health economics.

Thus many drugs can be adapted for sedation given adequate resources, i.e. time, anaesthetists, infusion pumps and recovery facilities. However, at present any practical discussion of sedation for endoscopic procedures has as its central theme the pharmacology and application of the benzodiazepines, specifically midazolam, its interactions with other drugs, principally the opioids, and the combined use of midazolam and flumazenil [10].

Benzodiazepines

History

The first benzodiazepine, chlordiazepoxide was synthesised in 1956 and introduced to clinical practice in 1960 as an orally administered anxiolytic. In 1977 specific benzodiazepine receptors were described in the central nervous system (CNS) [11, 12] associated with receptors for gamma-amino butyric acid (GABA) to form a protein complex containing a chloride ion channel referred to as the GABA-BZD-Cl$^-$ channel complex [13–18]. Similar receptors are now known to be widely distributed throughout the body [19] including the myocardium [20]. In the early 1980s midazolam Ro21-3965 (Dormicum®, Versed®, Hypnovel®, Doricum®, Dormonid®, Flormidal®) was introduced. It is the first water soluble benzodiazepine and has the shortest duration of action of all benzodiazepine agonists so far developed for clinical use. The first specific benzodiazepine antagonist flumazenil, Ro15-1788 (Anexate®, Lanexat®), which is also water soluble, was synthesised in 1979, characterised in 1981, and introduced into clinical prac-

Table 1. Reciprocal dose effect of benzodiazepine agonist and flumazenil

BZD agonist	Effect	Flumazenil
Low dose	Anxiolysis	High Dose
	Anticonvulsive effect	
	Slight sedation	
	Reduced attention	
	Amnesia	
	Intense sedation	
	Muscle relaxation	
High Dose	Hypnosis (anaesthesia)	Low dose

tice from the mid-1980s onwards. It introduced an agonist-antagonist technique whereby benzodiazepine-induced sedation, or anaesthesia, can be terminated at will [10].

Pharmacology

Receptors

Conventional benzodiazepines have full agonist effects at benzodiazepine receptors, causing, in increasing dose and receptor occupancy, anxiolysis through sedation, amnesia and muscle relaxation to anaesthesia (Table 1) although clearly there is considerable overlap of the various behavioural changes.

Ligands which bind to these receptors with the opposite effect to the benzodiazepines, to cause anxiety, arousal and convulsions are called *inverse agonists* (e.g. methyl β-carboline carboxylate, βCCM, [21–23] anxiety peptides [24–28]) while those which in themselves are not convulsive, but which kindle the system for other convulsive agents are referred to as *partial inverse agonists* (e.g. FG 7142, CGS 8216, ethyl β-carboline carboxylate – βCCE, Ro15-4513) [29]. In addition there are *partial agonists* (e.g. Ro16-6028, Ro17-1812, ZK91296, CGS-9896) [29] and *antagonists* (flumazenil, Ro15-3505, ZK93426). Thus there are five groups of ligands for benzodiazepine receptors [30] and flumazenil antagonises the other four groups. It reverses all the effects of conventional benzodiazepines without affecting their bioavailability or kinetics [31, 32]. At very high doses, (e.g. 75–100 mg.kg^{-1}), flumazenil shows some agonist activity in animal models.

The postulated receptor occupancies of benzodiazepine receptors for different effects are anticonvulsive 20–25%, anxiolytic 20–30%, sedative 25–50%, hypnotic 60–90% [33].

After midazolam, the reaction time, a measure of mental concentration, recovers at plasma concentrations between 100 and 25 ng.ml^{-1} [34], while sedation persists until the plasma concentration is below 75 ng.ml^{-1}. Amnesia is total above 100 ng.ml^{-1} [35] and anaesthesia occurs above 250 ng.ml^{-1} [36–38].

Table 2. Therapeutic index (ThI) = MT (maximum tolerated, i.e. highest non-lethal dose i.v. in rodents)/CD (clinical anaesthetic doses in man)

Drug	Mean MT mg.kg^{-1}	CD mg.kg^{-1}	ThI (man)
Midazolam	53.5	0.07–0.3	177+
Flumazenil	64.3	0.005–0.15	428+
Thiopentone	37.5	3.0–3.5	10.7+
Propofol	25	2.0–2.5	10+

Myocardial benzodiazepine receptors

The GABA-BZD-Cl$^-$ channel system has been demonstrated in myocardial papillary muscle of the guinea pig. It has the effect of decreasing myocardial contractility and shortening the duration of the action potential [20]. Benzodiazepines have been used in the treatment of arrhythmias associated with chloroquine poisoning [39].

General pharmacology

There are over 30 benzodiazepines on the market principally for oral administration. In general in suitable formulation they are absorbed by all routes, oral, sublingual, nasal, rectal, intramuscular and intravenous. The bioavailability after oral administration depends on first pass extraction by the liver and is around only 44% for midazolam and 16% for flumazenil which are the most rapidly metabolised benzodiazepines. Midazolam, diazepam, flunitrazepam, lorazepam and flumazenil are five commercially available drugs which can be administered parenterally.

Metabolism is by hydroxylation or decarboxylation, desalkylation (e.g. desmethyl, desethyl) followed by glucuronidation with both biliary and urinary excretion. The metabolites are often active and longer lasting than the parent drug so that the administered drug is a "pro-drug". Three, oxazepam, temazepam, lorazepam, are not metabolised, but merely glucuronated before excretion, and before the introduction of triazolam and midazolam were regarded as short acting (e.g. $t_{1/2}$ β for lorazepam ≈ 12 h).

In a sense benzodiazepines do not have primary effects but act through the GABA system. There is a ceiling to the effect they can produce which is dependent on receptor occupancy and ongoing GABA activity. They have a much higher therapeutic index than conventional intravenous induction agents, e.g. thiopentone, propofol (Table 2). In normal clinical doses their effects on the cardiovascular and respiratory systems are relatively small.

No anaphylactic reactions to benzodiazepines have been reported, apart from those in the early 1980s in Sweden due to a cremophor preparation of diazepam, which is no longer available. Futhermore, benzodiazepines can be used to induce anaesthesia in the presence of malignant hyperthermia and porphyria.

Table 3. Principal differences between midazolam and diazepam.

	Midazolam	Diazepam
Water solubility	+	diazepam emulsion
Pain on injection	< 2%	diazepam emulsion < 1%
		diazepam ≈ 40%
Metabolites	hydroxy (α 4-dihydro)	desmethyl
$t^{1}/_{2}\beta$ drug (hours)	2.5–3.5	30
$t^{1}/_{2}\beta$ metabolites (hours)	< 1.5	100
Enterohepatic		
circulation	–	+
Amnesia	++	+

Midazolam versus diazepam

The principal differences between midazolam and diazepam which can affect their clinical use are summarised in Table 3. These are primarily the pharmacokinetics where significant plasma concentrations of diazepam and its desmethyl derivative may persist for days with the potential for enterohepatic circulation, and hence midazolam is associated with earlier complete recovery and fewer longer term side effects than diazepam [40]. Midazolam also has the advantage of being water soluble and providing a greater degree of amnesia.

An equipotency ratio of 0.07 mg.kg^{-1} of midazolam to 0.15 mg.kg^{-1} of diazepam was proposed in the early 1980s [41]. It is now recognised that midazolam is more potent and that a relative dose of 0.05 mg.kg^{-1} or lower is more appropriate. Although the onset time of midazolam [41] is faster than diazepam the time to peak effect can be up to 5 minutes [42]. These two factors have probably combined so that in previous comparative studies [41] the dose of midazolam has been high relative to diazepam.

Bolus dose versus titration

Because of the slow onset times of the benzodiazepines it is possible for the most patient clinician to overdose with both drugs. By using a bolus injection of the benzodiazepine the sedative dose should be reduced by approximately one third for both midazolam [43] and diazepam [44] and some authorities use only approximately one half of the normal dose of midazolam and work in a dose range of 2–3 mg intravenously.

Patient controlled sedation (PCS)

A system for midazolam which has been evaluated by the author's group will be described. Midazolam 5 mg is diluted to 30 ml. The individual doses are 0.5 mg i.v., i.e. 3 ml delivered during 75 seconds, with a "lock out" time of 3 minutes and a maximum initial pre set dose of 5 mg in 1 hour.

Genetic polymorphism

In 1986, Dundee and his associates suggested, on the basis of retrospective data from studies conducted during the previous five to six years, that in about 6% of the population the terminal half life of midazolam was in the range 8–22 hours compared with the majority where the mean half life is about 2.4 hours. This would indicate polymorphism in the cytochrome P_{450} system for the metabolism of benzodiazepines and would have implications for the use of flumazenil. However Kassai in a prospective study on 168 hospitalised patients with normal hepatic and renal function found no evidence for this [46]. The mean plasma half lives for men and women were 2.6 and 3.1 hours respectively and in patients under and over 60 years 2.8 and 3.0 hours respectively [46]. Other factors which may prolong its half life are hypovolaemia [47], critical illness [48] and there may also be ethnic differences [49].

Drug interactions

Opioids and barbiturates

Cook and his associates described the use of intramuscular pethidine as a premedication 30–40 minutes before the induction of sedation for gastroscopy with intravenous diazepam [50]. Since then many gastroenterologists use not only pethidine but also a variety of other opioids intravenously during benzodiazepine sedation.

In addition to analgesia the opioids cause respiratory depression, bradycardia and hypotension due to baroreflex sensitisation [51], muscle rigidity and nausea and vomiting [52].

Ben-Shlomo and his associates [53] determined 50% of the effective dose (ED_{50}) for the induction of hypnosis with fentanyl and midazolam. They found that the administration of only 25% of the ED_{50} of fentanyl i.v. reduced the ED_{50} dose of midazolam by 77%, i.e. to 23% of its value when administered alone, showing genuine synergism between the two drugs.

Synergism of a similar order of magnitude has been shown to exist between midazolam and alfentanil [54], thiopentone [55], and methohexitone [56]. In contrast, midazolam and morphine have additive effects [57] and the position is summarised in Table 4.

Tucker and co-workers found that fentanyl caused a greater degree of respiratory depression with a significant reduction in oxygen saturation (SaO_2) when combined with both midazolam and diazepam [58].

The precise relationship between nalbuphine and the benzodiazepines has not been evaluated. Suffice it to say that nalbuphine prolongs the recovery time and the incidence of side effects of diazepam-induced sedation [59]. In one study two patients who received nalbuphine in addition to midazolam developed obstructive apnoea [60] in spite of the ceiling effect of nalbuphine on respiratory depression when administered alone [61].

Table 4. Midazolam drug interactions in man. ED_{50} of equi-effective doses for the induction of sleep or light anaesthesia as indicated by failure to open eyes on command. ED_{50} for each drug alone = 100% Numbers in table = percentage of the doses of each drug which when used in combination produce the same ED_{50} as each drug when used alone.

	Drug combination to produce ED_{50} of 100%			
	2nd Drug	Second Drug $ED_{50}\%$	Midazolam $ED_{50}\%$	Ref
Synergism				
Opioid	Fentanyl	25	23	[53]
	Alfentanil	33	21	[54]
Barbiturate	Thiopentone	25	25	[55]
	Methohexitone	33	27	[56]
Addition				
Morphine		50	50	[57]

The conclusion is that once a patient has received either midazolam or any other benzodiazepine, great care should be exercised in the administration of an opioid, e.g. fentanyl, alfentanil, nalbuphine, morphine and pethidine, or a barbiturate and vice versa. A major reduction in the dose of the second drug should be anticipated if the induction of anaesthesia with potentially dangerous side effects are to be avoided such as apnoea and loss of the airway, with subsequent hypoxaemia, and also a fall in arterial pressure. These remarks will also probably apply to a combination of many psychosedative drugs with the benzodiazepines.

H_2 receptor antagonists

The cytochrome P_{450} enzyme system is involved in the metabolism of the benzodiazepines in the liver and hence drugs such as cimetidine but possibly not ranitidine could be expected to prolong their action. However the position is not clear cut because in volunteer studies the concurrent administration of either oral cimetidine or ranitidine increased the bioavailability of oral midazolam by approximately 30% [62, 63]. Oral cimetidine caused a relatively small increase in the rate of fall in plasma levels of a sedative dose (5 mg i.v.) of this drug [64].

Ketorolac

Ketorolac [65] is a nonsteroidal anti-inflammatory drug (NSAID) with analgesic properties as good as morphine. However, it causes no cardiovascular or respiratory depression and an extremely small incidence of nausea. It warrants investigation as an analgesic supplement to sedative drugs such as midazolam. However, one of its side effects is somnolence and it could be that there will be an interaction in hypnotic effects.

Flumazenil

Flumazenil is a water-soluble benzodiazepine [33, 66–68]. It is a competitive antagonist at the benzodiazepine receptor and antagonises all other ligands for this receptor. It reverses all the modalities of effect of the benzodiazepines without affecting their bioavailability or kinetics and has minimal effects on the normal CNS with slight agonist and some anticonvulsive properties. Incremental doses up to 0.5–1.0 mg i.v. are effective clinically while doses of 600 mg orally and 100 mg i.v. are well tolerated in normal volunteers.

It is rapidly absorbed orally, with a high hepatic clearance and bioavailability averages only 16%. The mean volume of distribution (VD_{ss}) is 0.95 ± 0.16 $l.kg^{-1}$ (mean $\pm$ SD), clearance ($Cl_p d$) is 920 ± 216 $ml.min^{-1}$ and $t^{1}/_2 \beta$ is 0.9 ± 0.2 hours. Only 40% is bound to plasma proteins. It rapidly enters and subsequently clears the CNS. Less than 0.2% is excreted unchanged. Hydroxylation and desmethylation occur.

Its duration of action following a single injection i.v. has varied from 15–140 minutes depending on the dose but its mean duration of action in normal subjects is about 58 minutes.

In endoscopy clinics where conscious benzodiazepine sedation is used, flumazenil allows immediate reduction of sedation if verbal contact is lost or apnoea occurs. Because of its relatively short duration of action, to minimise the risk of resedation, long-acting agonist drugs should be avoided, and midazolam is the drug of choice. Flumazenil has been used to reverse midazolam-supplemented anaesthesia [69–71].

Flumazenil in clinical doses has similar side effects to placebo. Sixty seven out of 301 normal volunteers variously exhibited extremely transient episodes of dizziness, headache, flushing and anxiety following bolus intravenous doses in excess of 5 mg and as high as 60 mg [72].

Flumazenil reversal of sedation

There are now many studies reporting the use of flumazenil in reversing midazolam [73–76] and diazepam [73, 75, 77, 78] sedation particularly following endoscopy and minor surgery [75–79].

The potential indications of flumazenil in sedation are:
a) *Medical* (i.e. safety);
   i) as an emergency drug, e.g. in case verbal communication is lost and/or the breathing becomes obstructed and laboured or absent;
   ii) to deliberately reverse patients of poor physical status as soon as the procedure is complete.
b) *Health economics* – to speed up the flow of patients and to minimise dependence on fully staffed expensive conventional recovery facilities.

Although flumazenil reverses the prospective amnesic effects of benzodiazepines, that for the procedure is retained. Anxiety has been reported following the administration of flumazenil. Most such reactions are due to the rapid reversal of heavily sedated amnesic patients in an unfamiliar environment [73]. However, it can produce panic attacks in patients subject to such attacks [80].

Paradoxical reactions to the effects of midazolam in elderly patients, with anxiety and restlessness rather than sedation, have been successfully treated with flumazenil [73].

One of the major advantages of reversal of benzodiazepines with flumazenil, compared for example with the use of naloxone to reverse opioids, is that there is no cardiovascular rebound with hypertension [69, 81].

Previous areas of critical debate

These are resedation, the use of flumazenil in chronic benzodiazepine users and ventilatory responsiveness to changes in blood gas tensions.

Resedation

In the past the morality of the very existence of flumazenil has been challenged [82].

As the effects of flumazenil decline, i.e. about 45–60 min after its administration, the patient returns to the level of sedation which would have existed at that time if flumazenil had not been administered. Normally without flumazenil, given a correct dose of midazolam, a patient would be ready for discharge about one hour after completion of the procedure with no danger of resedation due to enterohepatic circulation, as is the case with diazepam, since the elimination of midazolam and its metabolites is complete within 5–8 hours in the type of patient undergoing outpatient gastroscopy (i.e. patients classified ASA I/II, controlled III – for details of ASA classification see Chapter 4, Table 1).

Thus the administration of flumazenil on completion of the procedure after sedation with midazolam provides, in terms of current practice, immediate clinical recovery since the patients can walk and sit in a chair, rather than recover on a trolley or bed, are no longer amnesic and are therefore capable of receiving instructions. After about one hour they can be accompanied home in the normal way with the advantage of not requiring conventional expensive recovery facilities in the immediate post-procedural period.

Midazolam is a highly potent drug and recent criticisms of the use of flumazenil to reverse its effects are due to a lack of understanding of the low dose of midazolam which is required to induce adequate clinical sedation. There also appears to be in some cases an ignorance of the pharmacology of the newer benzodiazepines since many clinicians regard midazolam as equivalent to diazepam in all respects. The current use of the original formulation of diazepam with its propensity to cause pain on injection and permanent venous thrombosis in countries where a diazepam emulsion preparation is available, would support the view that its continued use is to do with long standing prescribing habits, and arguably the continued use of either in preference to midazolam is sometimes based on similar considerations rather than on currently accepted pharmacological data.

Chronic benzodiazepine users

Concern has also been expressed that flumazenil could precipitate an acute withdrawal syndrome [83] in patients receiving prolonged benzodiazepine medication comparable to that induced by naloxone in chronic opioid users. Withdrawal symptoms, never prolonged or severe, have followed the administration of flumazenil in several species of benzodiazepine-treated animals [84]. However, a recent paper suggests that far from being a problem, flumazenil may be useful in the treatment of benzodiazepine tolerance [85]. This was an investigation using positron emission tomography (PET scanning) to study receptor occupancy in patients with temporal lobe epilepsy who had become tolerant to clonazepam. A single dose of flumazenil (1.5 mg i.v.) restored the effectiveness of clonazepam such that the patients were seizure-free for 6-21 days. They attributed the absence of withdrawal seizures to three factors:
a) anticonvulsive effects of flumazenil;
b) the fact that 1.5 mg of flumazenil only occupies 45–55% of the benzodiazepine receptors and clonazepam 2–5 mg only 30%;
c) configurational "resetting" of the receptors by flumazenil.

An intriguing concept which could follow from this work is that patients on long-term benzodiazepine administration, who could have become tolerant to the effects of benzodiazepines and hence have an unpredictable response for example to midazolam, should receive 2–3 mg of flumazenil to reset say over 60% of the receptors. Approximately one-two hours later, when the antagonistic effect of the flumazenil has disappeared, the patient may then show a normal response to midazolam which could subsequently be reversed by flumazenil in the normal way.

Ventilatory response to hypoxaemia and hypercarbia

It has been suggested that whereas the effect of flumazenil in reversing sedation is immediate, full recovery of chemoreflexes is delayed [86, 87]. However, there are several problems in the interpretation of recent studies on the effect of midazolam on the ventilatory responses to hypoxaemia and carbon dioxide.
a) Normal subjects show a large random variation in responsiveness over time and on different days [88]. Power and associates in controlled studies, where a placebo had been administered, lasting over six hours found spontaneous variations of the standard deviations of responses in

normal volunteers in excess of 50% of the mean values for the slope of the CO_2 response curve and over 25% for the intercept [88]. Thus changes less than this in response to a drug, even if statistically significant, would not have any practical significance. Thus when using CO_2 response curves as a method of evaluating respiratory depression, it is not enough to undertake a control reading. Rather, the normal variation in the group over time or on different days, must be investigated before any conclusions can be drawn about the depressant effect of psychosedative drugs unless these are obvious or verge on apnoea as can be the case with the opioids and barbiturates.

b) Responses to hypoxaemia are non-linear.

c) During a period of sustained hypoxaemia as can happen during gastroscopy central adaptations occur so that the initial vigorous response, which is not affected by small doses of midazolam, declines [89] which could be falsely interpreted as due to the depressant effect of the sedative drug.

d) Earlier studies on the effects of midazolam and diazepam showed that neither had any great effect on CO_2 responsiveness. For example, Forster and associates found that although 0.15 $mg.kg^{-1}$ of midazolam and 0.3 $mg.kg^{-1}$ of diazepam caused a reduction in the slope of the CO_2 response curve at 4 mins there was no change in the intercept [90]. Moreover Power and associates showed that neither midazolam 0.075 $mg.kg^{-1}$ nor diazepam 0.15 $mg.kg^{-1}$ caused any significant changes in CO_2 responsiveness [91]. In general benzodiazepines cause minimal respiratory depression [92].

e) It has also been suggested that during prolonged sedation with diazepam, mean duration 145 min (dose 1 $mg.min^{-1}$ i.v.) that the recovery of ventilatory depression is due to the onset of tolerance to diazepam [87], a phenomenon which has been described previously for midazolam in animals [93].

f) Recent evidence suggests that 15 mg flumazenil may be required in man for full receptor occupancy [85]. Thus the suggested failure of up to 2.0 mg of flumazenil to reverse completely all the effects of the benzodiazepines may be simply a question of inadequate dosage. Moreover, if an element of acute tolerance, for example to midazolam, has been induced by prolonged sedation which can occur within 2 hours in animals [93], and has been postulated by Mora *et al* for diazepam in man [87], then it has been shown in man that flumazenil could reverse such tolerance e.g. to clonazepam [85]. This would restore the effectiveness of the agonist which would then require more flumazenil to block the renewed activity.

Until the position is clarified by correctly designed relevant studies, pulse oximetry should be used to monitor all patients undergoing benzodiazepine sedation, and they should be observed until the control SaO_2 returns when breathing air.

Flumazenil is specific for the benzodiazepine component of sedation. The effect of a second type of drug e.g. an opioid, will persist [94].

Conclusions

The pharmacological properties of midazolam make it the benzodiazepine of choice for sedation for procedures and in outpatients where rapid recovery is desirable. Although flumazenil is a relatively short-acting drug, when its effects decline the patient will return to the point of residual sedation that would have been present at that time [74]. Thus although flumazenil can facilitate the management of patients during the first hour after its administration, thereafter the instructions to the patient and accompanying person for day case or ambulatory surgery are the same whether they have received flumazenil or not. Hence the most important factor in safety is either to titrate or select the initial dose of midazolam so that the patient receives a suitable amount of the drug and is not oversedated. The sensitivity of the elderly to midazolam and their slower circulation times should be considered when titrating the minimal effective dose. Monitoring with pulse oximetry is mandatory. Safety demands trained staff with all necessary equipment and drugs which are regularly either checked for expiry date or serviced (e.g. battery replacement).

The introduction of flumazenil is a major advance enhancing the safety of benzodiazepine-induced sedation. However, it will only reverse the benzodiazepine component of sedation and for example where an opioid has also been used its effects will persist.

A single dose of flumazenil will only facilitate the management of patients during the first 45–60 min of the recovery period. Thereafter the same rules must apply for benzodiazepine sedation as would be the case if flumazenil had not been administered.

Safety considerations apart, the correct use of midazolam and post-procedural administration of flumazenil could lead to a more efficient transit of patients allowing one aspect of the medical needs of society to be satisfied within both the financial and human resources, particularly nurses, available to treat them.

References

1. Birch BRP, Anson KM, Miller RA (1990) Sedoanalgesia in urology: a safe, cost-effective alternative to general anaesthesia. A review of 1020 cases. Br J Urol 66:342–350
2. Department of Health Standing Dental Advisory Commitee (1991) Report of an expert working party. General anaesthesia, sedation and resuscitation in dentistry (In press)
3. Cockshott ID (1985) Propofol (Diprivan). Pharmacokinetics and metabolism – an overview. Postgrad Med J 61 (Suppl 3):45–50
4. Mackenzie N, Grant IS (1987) Propofol for intravenous sedation. Anaesthesia 42:3–6

5. Roberts FL, Dixon J, Lewis GTR, Tackley RM, Prys-Roberts C (1988) Induction and maintenance of propofol anaesthesia. A manual infusion scheme. Anaesthesia 43 (Suppl):14–17

6. Rudkin GE, Osborne GA, Curtis NJ (1991) Intra-operative patient-controlled sedation. Anaesthesia 46:90–92

7. Thompson GE, Moore DC (1971) Ketamine, Diazepam and Innovar. A computerised comparative study. Anesth Analg 50:458–463

8. Podlesch I, Dähn H, Engel M (1985) Anaesthetic and side effects of midazolam and ketamine in surgical patients. Anesthesiology 62:68

9. Seitz W, Lübbe N, Hamkens A, Bornscheuer A (1988) Midazolam-Ketamin-Kombinationsanaesthesie bei traumatologischen Eingriffen. Anaesthetist 37:231–237

10. Thomson D (ed) (1990) Midazolam and Flumazenil – the agonist-antagonist concept for sedation and anaesthesia. Proceedings of an international symposium, April 1989. Acta Anaesthesiol Scand 34 (Suppl 92) 1–109

11. Mohler H, Okada T (1977) Benzodiazepine receptor: demonstration in the central nervous system. Science 198:849–851

12. Braestrup C, Squires RF (1977) Specific benzodiazepine receptors in rat brain characterised by high-affinity ^{3}H-diazepam binding. Proceedings of the National Academy of Sciences USA 74:3805–3809

13. Richards JG, Möhler H (1984) Benzodiazepine receptors. Neuropharmacology 23:233–242

14. Braestrup C, Nielsen M (1982) Neurotransmitters and CNS disease; anxiety. Lancet 2:1030–1034

15. Chang LR, Barnard EA (1982) The benzodiazepine/GABA receptor complex: molecular size in brain synaptic membranes and in solution. J Neurochem 39:1507–1518

16. Schoch P, Richards JG, Häring P, Takacs B, Stähli C, Staehelin T, Haefely W, Möhler H (1985) Co-localisation of GABA and benzodiazepine receptors in the brain. Nature 314:168–170

17. Möhler H, Schoch P, Richards JG, Häring P, Takacs B, Stähli C (1986) Monoclonal antibodies: probes for structure and location of the GABA receptor/benzodiazepine receptor/chloride channel complex. In: Benzodiazepine-GABA receptors and chloride channels: structural and functional properties. Olsen RW, Venter JC (eds). New York: Alan R Liss, Inc.

18. Schofield PR, Darlinson MG, Fujita N, Burt DR, Stephenson FA, Rodriguez H, Rhee LM, Ramachandran J, Reale V, Glencorse TA, Seeburg PH, Barnard EA (1987) Sequence and functional expression of the GABA receptor shows a ligand-gated receptor super-family. Nature 328:221–227

19. Erdö S (1985) Peripheral GABAergic mechanisms. Trends in Pharmacological Sciences 6:205–208

20. Mestre M, Carriot T, Belin C, Uzan A, Renault C, Dubroeqcq MC, Gueremy C, Le Fur G (1985) Electrophysiological and pharmacological characterization of peripheral benzodiazepine receptors in a guinea pig heart preparation. Life Sci 35:953–962

21. Braestrup C, Nielsen M, Olsen CE (1980) Urinary and brain β-carboline-3-carboxylates as potent inhibitors of brain benzodiazepine receptors. Proc Natl Acad Sci USA 77:2288–2292

22. Braestrup C, Nielsen M (1981) GABA reduced binding ^{3}H-methyl β-carboline-carboxylate to brain benzodiazepine receptors. Nature 294:472–475

23. Braestrup C, Schiechen R, Neef G, Nielsen M, Petersen EN (1982) Interaction of convulsive ligands with benzodiazepine receptors. Science 216:1241–1243

24. Guidotti A, Forchetti CM, Corda MG, Konkel D, Bennett CD, Costa E (1983) Isolation, characterization and purification to homogeneity of an endogenous polypeptide with agonist action on benzodiazepine receptors. Proc Natl Acad Sci USA 80:3531–3535

25. Editorial (1987) Diazepam binding inhibitor. Lancet 1:307–308
26. Ferrero P, Sahti MR, Conti-Tronconi B, Costa E, Guidotti A (1985) Study of an octadecaneuropeptide derived diazepam binding inhibitor (DBI): biological activity and presence in rat brain. Proc Natl Acad Sci USA 83:827–831
27. Marquardt H, Todaro GJ, Shoyab M (1986) Complete amino acid sequences of bovine and human endozepines. J Biol Chem 261:9727–9731
28. Ferrero P, Guidotti A, Conti-Troncon B, Costa E (1984) A brain octadecaneuropeptide generated by tryptic digestion of DBI (diazepam binding inhibitor) functions as a proconflict ligand of benzodiazepine recognition sites. Neuropharmacology 23:1359–1362
29. Little HJ, Nutt DJ, Taylor SC (1987) Kindling and withdrawal changes at the benzodiazepine receptor. Journal of Psychopharmacology 1:35–46
30. Richards JG, Schoch PO, Möhler H, Haefely W (1986) Benzodiazepine receptors resolved. Experientia 42:121–126
31. Darragh A, Lambe R, Kenny M, Brick I, O'Boyle C, Taffe W (1982) Ro15–1788 antagonises the central diazepam in man without altering diazepam availability. Br J Clin Pharmacol 14:677–682
32. O'Boyle C, Lambe R, Darragh A, Taffe W, Brick I, Kenny M (1983) Ro15–1788 antagonises the effects of diazepam in man without affecting its bioavailability. Br J Anaesth 55:349–355
33. Amrein R, Hetzel W (1990) Pharmacology of Dormicum (midazolam) and Anexate (Flumazenil). Acta Anaesthesiol Scand (Suppl 92) 34:6–15
34. Crevoisier C, Ziegler WH, Heizmann P, Dubuis R (1984) Relation entre l'effet clinique et la concentration plasmatique du midazolam chez des sujets volontaires. Ann Fr Anesth Reanim 3:162–167
35. Persson MP, Nilsson A, Hartvig P (1988) Relation of sedation and amnesia to plasma concentrations of midazolam in surgical patients. Clin Pharmacol Ther 43:324– 331
36. Lauven PM, Stoeckel H (1987) Hypnotische wirksame Blutspiegel von Midazolam. Anasth Intensivther Notfallmed 22:90–93
37. Raeder JC, Nilsen G, Hole A (1988) Pharmacokinetics of midazolam and alfentanil in outpatient general anaesthesia. Acta Anaesthesiol Scand 32:467–472
38. Lauven PM, Schwilden H, Stoeckel H, Greenblatt DJ (1985) The effects of a benzodiazepine antagonist Ro15–1788 in the presence of stable concentrations of midazolam. Anesthesiology 63:61–64
39. Riou B, Rimailho A, Galliot M, Bourdon R, Huet Y (1988) Protective cardiovascular effects of diazepam in chloroquine poisoning. Intensive Care Med 14:610–616
40. Sanders LD, Davies-Evans J, Rosen M, Robinson JO (1989) Comparison of diazepam with midazolam as i.v. sedation for outpatient gastroscopy. Br J Anaesth 63:726–731
41. Whitwam JG, Al-Khudhairi D, McCloy RF (1983) Comparison of midazolam and diazepam in doses of comparable potency during gastroscopy. Br J Anaesth 55:773–777
42. Gamble JAS, Kawar P, Dundee JW, Moore J, Briggs LP (1981) Evaluation of midazolam as an intravenous induction agent. Anaesthesia 36:868–873
43. Bell GD, Antrobus JHL, Lee J, Coady T, Morden A (1990) Bolus or slow titrated injection of midazolam prior to upper gastrointestinal endoscopy? Relative effect on oxygen saturation and prophylactic value of supplemental oxygen. Alimentary Pharmacology and Therapeutics 4:393–401
44. Swain DG, Ellis DJ, Bradby H (1990) Rapid intravenous low-dose diazepam as sedation for upper gastrointestinal endoscopy. Alimentary Pharmacology and Therapeutics 4:43–48
45. Dundee JW, Collier PS, Carlisle RJT, Harper KW (1986) Prolonged midazolam elimination half-life. Br J Clin Pharmacol 21:425–429
46. Kassai A, Toth G, Eichelbaum M, Klotz U (1988) No evidence of a genetic polymorphism in the oxidative metabolism of midazolam. Clin Pharmacokinet 15:319–325

47. Adams P, Gelman S, Reves JG, Alvis M, Greenblatt DT et al (1985) Midazolam pharmacokinetics and pharmacodynamics during acute hypovolemia. Anesthesiology 63:140–146
48. Shelly MP, Mendel L, Park GR (1987) Failure of critically ill patients to metabolise midazolam. Anaesthesia 42:619–626
49. Kalow W (1982) Ethnic differences in drug metabolism. Clin Pharmacokinet 7:373–400
50. Cook PJ, Bennett PN, Lennard-Jones JE, Warnes TW (1978) Premedication for endoscopy. Scand J Gastroenterol 13:33–39
51. Swenzen GO, Chakrabarti MK, Sapsed-Byrne S, Whitwam JG (1988) Selective depression by alfentanil of Group III and IV somatosympathetic reflexes in the dog. Br J Anaesth 61:441–445
52. Whitwam JG (1988) Summary of meeting. In:Kallar SK, Whitwam JG (eds). Outpatient anaesthesia. Proceedings of an international symposium, Antwerp, Belgium, 9th June 1988 Amsterdam:Medicom
53. Ben-Shlomo I, Abd-El-Khalim H, Ezry J, Zohar S, Tverskoy M (1990) Midazolam acts synergistically with fentanyl for induction of anaesthesia. Br J Anaesth 64:45–47
54. Vinik HR, Bradley EL, Kissin I (1989) Midazolam-alfentanil synergism for anaesthetic induction in patients. Anesth Analg 69:213–217
55. Tverskoy M, Fleyshman G, Bradley EL, Kissin I (1988) Midzolam-thiopental anesthetic interaction in patients. Anesth Analg 67:342–345
56. Tverskoy M, Ben-Shlomo I, Ezry J, Finger J, Fleyshman G (1989) Midazolam acts synergistically with methohexitone for induction of anaesthesia. Br J Anaesth 63:109–122
57. Tverskoy M, Fleyshman G, Ezry J, Bradley EL, Kissin I (1989)Midzolam-morphine sedative interaction in patients. Anesth Analg 68:282–285
58. Tucker MR, Ochs MW, White PW (1986) Arterial blood gas levels after midazolam or diazepam administered with or without fentanyl as an intravenous sedative for outpatient surgical procedures. J Oral Maxillofac Surg 44:688–692
59. Dolan EA, Murray WJ, Immediata AR, Gleason N (1988) Comparison of nalbuphine and fentanyl in combination with diazepam for outpatient oral surgery. J Oral Maxillofac Surg 46:471–473
60. Sury MRJ, Cole PV (1988) Nalbuphine combined with midazolam for outpatient sedation. An assessment in fibreoptic bronchoscopy patients. Anaesthesia 43:285–288
61. Romagnoli A, Keats AS (1980) Ceiling effect for respiratory depression by nalbuphine. Clin Pharmacol Ther 27:478–485
62. Elwood RJ, Hildebrand PJ, Dundee JW, Collier PS (1983) Ranitidine influences the uptake of oral midazolam. Br J Clin Pharmacol 15:743–745
63. Fee JPH, Collier PS, Howard PJ, Dundee JW (1987) Cimetidine and ranitidine increase midazolam bioavailability. Clin Pharmacol Ther 41:80–84
64. Knüchel M, Ochs HR, Verbrug-Ochs B, Labedzki L, Greenblatt DJ (1987) Interaktion von Ranitidin und Cimetidin mit Midazolam bei intravenöser und oraler Gabe. Med Welt 38:244–248
65. Buckley MMT, Brogden RM (1990) Ketorolac. A review of the pharmacodynamic and pharmacokinetic properties and therapeutic potential. Drugs 39:86–109
66. Amrein R, Leishman B, Bentzinger C, Roncari G (1987) Flumazenil in benzodiazepine antagonism. Actions and clinical use in intoxication and anaesthesiology. Medical Toxicology 2:411–429
67. Klotz U, Kanto J (1988) Pharmacokinetics and clinical use of flumazenil. Clin Pharmacokinet 14:1–12
68. Whitwam JG (1988) Flumazenil: a benzodiazepine antagonist. Br Med J 297:999–1000

69. Fisher GC, Hutton P (1989) Cardiovascular responses to flumazenil-induced arousal after arterial surgery. Anaesthesia 44:104–106
70. Wolff J, Carl P, Clausen TG, Mikkelsen BO (1986) Ro15–1788 for postoperative recovery. A randomised clinical trial in patients undergoing minor surgical procedures under midazolam anaesthesia. Anaesthesia 41:1001–1006
71. Alon E, Baitella L, Hossli G (1987) Double-blind study of the reversal of midazolam-supplemented general anaesthesia with Ro15–1788. Br J Anaesth 59:455–458
72. Data on file. F Hoffmann-La Roche Ltd., Basle, Switzerland
73. Ricou R, Forster A, Bruckner A, Chastonay P, Gemperle M (1986) Clinical evaluation of a specific benzodiazepine antagonist (Ro 15–1788). Studies in elderly patients after regional anaesthesia under benzodiazepine sedation. Br J Anaesth 58:1005–1011
74. Sage DJ, Chase A, Boas RA (1987) Reversal of midazolam sedation with Anexate. Br J Anaesth 59:459–464
75. Jensen S, Knudsen L, Kirkegaard L (1988) Flumazenil used in the antagonizing of diazepam and midazolam sedation in outpatients undergoing gastroscopy. Eur J Anaesthesiol Suppl 2:161–166
76. Rodrigo C, Rosenquist JB (1987) The effect of Ro15–1788 (Anexate) on conscious sedation produced with midazolam. Anaesth Intensive Care 15:185–192
77. Holloway AM, Logan DA (1988) The use of flumazenil to reverse diazepam sedation after endoscopy. Eur J Anaesthesiol Suppl 2:191–198
78. Kirkegaard I, Knudsen L, Jensen S, Kruse A (1986) Benzodiazepine antagonist Ro15–1788. Antagonism of diazepam sedation in outpatients undergoing gastroscopy. Anaesthesia 41:1184–1188
79. Andrews PJD, Wright DJ, Lamont MC (1988) Flumazenil in the outpatient. A study following midazolam as sedation for upper gastrointestinal endoscopy of sedation and amnesia to plasma concentrations of midazolam in surgical patients. Clin Pharmacol Ther 43:324–331
80. Nutt D, Glue P, Lawson C, Wilson S (1990) Flumazenil provocation of panic attacks. Arch Gen Psychiatry 47:917–925
81. Louis M, Forster A, Suter PM, Gemperle M (1984) Clinical and haemodynamic effects of a specific benzodiazepine antagonist (Ro 15–1788) after open heart surgery. Anesthesiology 513A
82. Editorial (1988) Midazolam – is antagonism justified? Lancet 1:140–142
83. Ashton H (1984) Benzodiazepine withdrawal: an unfinished story. Br Med J 288:1135–1140
84. Cumin R, Bonetti EP, Scherschlicht R, Haefely WE (1982) Use of the specific benzodiazepine antagonist Ro15–1788 in studies of physiological dependence on benzodiazepine. Experientia 38:833–834
85. Savic I, Widen L, Stone-Elander S (1991) Feasibility of reversing benzodiazepine tolerance with flumazenil. Lancet 337:133–137
86. Dell GD, Reeve PA, Moshiri M et al (1987) Intravenous midazolam: a study of the degree of oxygen desaturation occurring during upper-gastrointestinal endoscopy. Br J Clin Pharmacol 23:703–708
87. Mora CT, Torjman M, White PF (1989) Effects of diazepam and flumazenil on sedation and hypoxic ventilatory response. Anesth Analg 68:473–478
88. Power SJ, Chakrabarti MK, Whitwam JG (1984) Response to carbon dioxide after oral midazolam and pentobarbitone. Anaesthesia 39:1183–1187
89. Deham A, Ward DS (1991) Effect of i.v. midazolam on the ventilatory response to substained hypoxia in man. Br J Anaesth 66:454–457
90. Forster A, Fardaz JP, Suter PM, Gemperle M (1980) Respiratory depression by midazolam and diazepam. Anesthesiology 53:494–497
91. Power SJ, Morgan M, Chakrabarti MK (1983) Carbon dioxide response curves following midazolam and diazepam. Br J Anaesth 55:837–841
92. Dundee JW, Wyant GM (1974) Intravenous anaesthesia. Edinburgh: Churchill Livingstone

93. Al-Khudhairi D, Askitopoulou H, Whitwam JG (1982) Acute tolerance to the central respiratory effects of midazolam in the dog. Br J Anaesth 54:953–958
94. Weinbrum A, Geller E (1990) The respiratory effects of reversing midazolam sedation with flumazenil in the presence or absence of narcotics. Acta Anaesthesiol Scand Suppl 92 34:65–69

CHAPTER 2

Report of Workshop on Drugs for Sedation

Professor ERAN GELLER

Prevalence of sedation techniques

In the American survey conducted by Keeffe [1] it was shown that most patients routinely receive intravenous sedation for endoscopic procedures. Likewise, in the United Kingdom, Bell [2] has shown that approximately 90% of endoscopists use intravenous benzodiazepines for sedating at least 75% of patients undergoing endoscopy. Overall, the Workshop concurred with this view although there were some marked national variations. The most extreme was Finland, where Dr. Janatuinen reported that they do not use sedation at all for endoscopy except for children under 6 years old.

Use of drugs

In the United States the benzodiazepine midazolam is the drug used most frequently for sedation in endoscopy, whilst in the United Kingdom, diazepam is still being used more frequently. Professor Whitwam commented that the pharmacokinetic profiles of the two drugs (see u. c. chapter 1 pages 8, 9) clearly make the use of diazepam illogical for short endoscopic procedures. Most of the Workshop members agreed that intravenous midazolam has the attributes that make it the benzodiazepine of choice to use for sedation in endoscopy.

Sedation end points

McCloy and Pearson [3] have suggested that the end points to be avoided during intravenous sedation are ptosis, dysarthria and drowsiness (Table 1). These end points are dangerously close to a state of anaesthesia. Instead,

Table 1. Aims of sedation

No	Yes
ptosis	anxiolysis
dysarthria	amnesia
drowsiness	cooperation

the aim of intravenous sedation should be to relieve anxiety, induce amnesia and maintain cooperation with the patient. More research into how to achieve this state consistently is needed. Trieger in 1989 [4] suggested a clinical sign for appropriate sedation during oral surgery. It may also be considered as a practical end point during endoscopy, namely watching the patient's posture becoming relaxed, and while lying on the examination table the toes point outwards and the patient will say that he feels lightheaded. This is perhaps a practical way of judging the correct clinical end point of the desired state of sedation.

Methods of administering benzodiazepines

Bolus administration versus titration

In the United Kingdom survey [2] about one third of the respondents reported that they use a bolus technique of administering sedatives rather than titrating to an end point. An important question is whether endoscopists really titrate a drug, since few probably ever wait the 3–5 minutes required for each dose of the intravenous benzodiazepine to reach its peak sedative effect. Only two published studies [5, 6] address the issue of bolus versus titration (Table 2). Onset of sedation was found to be faster when the drug was given as a bolus over 5–10 seconds but the dose had to be reduced by about one third. Patient comfort was similar with both techniques. The risk of a significant fall in oxygen saturation or respiratory depression was increased in the bolus group especially if the bolus dose was inappropriate for a particular patient [5]. The use of supplemental oxygen should always be considered. Swain [6] reported that one patient had severe respiratory depression after bolus dosing. The total dose of diazepam in Swain's study was larger in the titration group and therefore more patients in this group were amnesic. Patients recovered faster in the group that received the drug

Table 2. Bolus versus titration

	Speed of onset	Coope-ration	Red. of O_2 Sat/ Resp. Depression	Total dose	Amnesia	Recovery
Titration	–	+	–	+	+	–
Bolus	+	+	+ (1 pt)	–	–	+
For Bolus Administration Recommended:		2/3 – 1/2 titrated dose Adapt dose to age Avoid opioids Avoid in sick patients Use supplemental O_2				

Bell et al 1990: Swain et al 1990

in a bolus, probably because they received a smaller dose. The Workshop members recommended that when using a bolus technique the dose of the benzodiazepine should be reduced to about one half, or even less, of the dose that is appropriate for a titration technique. The dose of the benzodiazepine should also be modified according to the age of the patient [7]. The bolus technique should be avoided in patients receiving opioids, because of the interaction between the drugs, (see below) as well as in patients who are critically ill, hypovolaemic or with chronic lung disease.

Alternative routes of administration

Langendijk from Amsterdam introduced the issue of the rectal application of midazolam. He reported on a study [8] in volunteers comparing intravenous and rectal administration of midazolam. The bioavailability per rectum is about 40% that of the intravenous route so that double the dose has to be given rectally. After 20 minutes comparable blood levels with rectal and intravenous methods are achieved and from that point on the pharmacokinetics of the drug are the same irrespective of the route of administration. Rectal midazolam has already been tested widely in children [9] as a premedication for anaesthesia and it may also have a place in endoscopy. Another point that was discussed briefly was the application of midazolam intranasally. This has also been studies in children undergoing anaesthesia in France [10] and to a lesser extent in the United States [11]. It may be interesting to evaluate this technique in children and adults undergoing endoscopy.

Combination of benzodiazepines and opioids

In the American survey [1] it was shown that, during 1989–1990 about 87% of endoscopists used an opioid, most commonly meperidine (pethidine) in addition to a benzodiazepine. In the United Kingdom survey only 13% of respondents [2] reported that they used meperidine (pethidine) in combination with one of the benzodiazepines. Interestingly, only a small proportion of endoscopists are using the newer opioids, such as fentanyl or alfentanil, which from a pharmacokinetic point of view may be more appropriate than the longer acting meperidine. The Workshop members recommended that in general polypharmacy should be avoided – a view supported by the literature. Not only is there a synergistic effect of the benzodiazepine/opioid combination on respiratory depression but also loss of verbal contact and unconsciousness are reached much faster.

In discussion, Professor Fleischer emphasised that the combination of a benzodiazepine with an opioid has been standard practice in the USA for a great many years. Therefore, are the dangers of synergism more theoretical than apparent? Professor Whitwam replied that there have been several cases resulting in litigation in the USA due to the alleged misuse of the combination. However, if you use a carefully standardised technique based on a strict protocol which has been evaluated for safety, the drug interactions will have been taken into account. Professor Geller commented that one reason why endoscopists may not be having problems with this drug interaction is because, at the same time as giving these two depressant drugs, the endoscopist is stimulating the patient very strongly by the insertion of the endoscope. It is interesting to speculate what would happen if an endoscopist was called to the telephone after injection of boluses of the opioid and the benzodiazepine and five minutes elapsed before starting the endoscopy. The chances are that the patient would have developed severe respiratory depression.

If it is necessary to use a combination of a narcotic together with a benzodiazepine the opioid should be injected first, since this is the major cause of severe respiratory depression. It tends to be the first dose of a benzodiazepine which causes respiratory depression – further increments do not usually compound the problem. This is not true for the opioids. The more opioid given, the greater the respiratory depression. If a benzodiazepine has already been administered and additional doses of the opioid are titrated there is an increased risk of severe respiratory depression.

The safest approach to combining benzodiazepine and opioids, if one chooses to do so, is to first administer a smaller dose of the opioid – say 25–30% of the usual dose. Once the opioid has produced its effect, which may take two to five minutes if given intravenously, or 20–30 minutes intramuscularly, the benzodiazepine should be carefully titrated using only about 25% of the dose when it is used alone. This technique allows for the synergism between the drugs. (see Chapter 1, reference 53). Additional doses of the benzodiazepine can be titrated if necessary.

Dr. Miller, a minimally invasive urologist, considered it important to employ an analgesic for therapeutic endoscopy or major endoscopic surgery. He routinely uses a nonsteroidal anti-inflammatory drug, diclofenac suppositories, prior to all endoscopic surgery, and has found this to be very successful. Midazolam is then given as an intramuscular premedication, in a dose according to the patient's weight, about 20 minutes before the procedure begins. The patient then arrives in the theatre appropriately sedated. Using traditional intravenous techniques many gastrointestinal endoscopists may start the procedure before the benzodiazepine and opioid have had time to reach their peak effects.

Professor Arnold made the point that pain can be a useful indicator, particularly during colonoscopy, for example, if the bowel is being unduly distended with the risk of perforation, and excessive analgesia may mask this.

Abnormal reactions to benzodiazepines

There was a consensus in the Workshop group that abnormal reactions to benzodiazepines are rare but when they occur they can be disturbing. Two

Table 3. Abnormal Reactions to Benzodiazepines

Type 1   paradoxical excitation
        (drug idiosyncrasy)

Treatment

● Flumazenil titration

Type 2   uncooperative and restless (?hypoxaemia? oversedation)

Treatment options

● supplemental oxygen
● flumazenil titration
  second drug e.g. droperidol or opioid
  (watch for synergism)
● call anaesthesiologist (for deep sedation/general anaesthesia)
● fully reverse with flumazenil (abandon procedure and re-evaluate the clinical situation)

---

types of reaction were identified in the Workshop discussions (Table 3). The first type is characterised by excitation, agitation and restlessness shortly following the injection of the benzodiazepine. This rare paradoxical reaction has been reported in the anaesthetic literature and is easily reversed with flumazenil which abolishes the excitability [12]. Although this has not been formally tested, Professor Geller expressed the view that if flumazenil is slowly titrated it may be possible to retain anxiolysis and light sedation while abolishing the excitability.

The second more common type of reaction occurs when the patient becomes uncooperative and restless. What is the cause and what should be done? Again this is an infrequent occurrence but it was raised by some of the Workshop participants as occasionally being a problem. It was agreed that this is usually a problem of oversedation and/or hypoxaemia. There are several ways of dealing with this problem (Table 3). Certainly, additional doses of benzodiazepine should not be added. Supplemental oxygen should be given and adequacy of breathing and arterial oxygenation should be checked. One can also titrate incremental doses of flumazenil to reverse oversedation and reach the point where the patient can control his reactions and cooperate. If hypoxaemia is not the problem, it might be necessary to add a second drug, for example droperidol or an opioid and such combinations can be dangerous (see above). Furthermore, as stated in the Poswillo Report [13], when more than one centrally acting drug is administered, this is no longer conscious sedation, and may be regarded medico-legally as anaesthesia. The practicality of calling an anaesthesiologist in a very busy endoscopy unit was ruled out by many of the Workshop participants. The last possibility would be to use flumazenil to reverse fully the patient's sedation and reschedule the procedure.

Use of flumazenil

Routine use of flumazenil

It was generally agreed that flumazenil should not be used to enable heavy sedation with large doses of agonists.

Many of the published studies [14–17] on reversal of benzodiazepine sedation in endoscopy involving several hundred patients conclude that reversal of sedation with flumazenil will improve patient throughput and allow earlier discharge of the patients.

Some centres, notably those with scarce recovery facilities and those practising more deep sedation for colonoscopies and ERCPs have used flumazenil effectively to increase workloads. However, although a case can be made for routine use of flumazenil in some endoscopic procedures more detailed safety and cost-benefit studies are needed to address this important issue.

Selective reversal

Flumazenil could be used selectively in high risk patients such as critically ill, or hypovolaemic patients and those with respiratory or cardiovascular disease. The aim would be to reverse sedation at the end of endoscopy to have a patient in the same condition as before the administration of the benzodiazepine.

During endoscopy, the stimulation caused by the procedure will counteract the depressive effects of sedation. However, when debilitated patients, known to be more sensitive to sedative drugs, are moved to the recovery area with no further stimulation, they can become more deeply sedated. Thus reversing sedation with flumazenil may increase patients' safety during recovery and diminish the burden on the nursing staff.

Dose of flumazenil

Limited information is available in the literature concerning the dose response relationship for efficacy and duration of action of flumazenil versus various agonists. Dunton [19] studied the efficacy and duration of effects of 0.2, 0.6, 1 and 3 mg of flumazenil in volunteers heavily sedated with a continuous infusion of midazolam. The mean duration of reversal of sedation with 3 mg flumazenil was more than double that achieved with the lower doses (103 versus approximately 46 min). Efficacy of reversal was similar with doses in the range of 0.6–3 mg while 0.2 mg was much less effective.

Further studies, under various clinical conditions, are necessary in order to define the exact relationships between different agonists and the antagonist.

Resedation

The issue of resedation has been recently reviewed by Whitwam [18]. When clinically correct doses of benzodiazepines are used no resedation occurs. In none of the studies on the use of flumazenil to reverse benzodiazepine sedation in endoscopy has resedation been reported to be a problem.

In the Workshop, Dr. Armegol Miro from Spain presented his experience using a long acting benzodiazepine, flunitrazepam, for endoscopy. He reported on about 1000 endoscopies that he had performed within the last 2 years using flunitrazepam for sedation and reversing all patients with flumazenil. In this series two patients developed severe respiratory depression which was immediately reversed by flumazenil. He did not observe any resedation. The panelists felt that short acting benzodiazepines, like midazolam, are preferred for sedation in endoscopy.

Chronic benzodiazepine users

Concern has been raised about the likelihood of flumazenil precipitating withdrawal reaction in chronic benzodiazepine users.

Dunton [19] has found no signs of withdrawal after 1 mg of flumazenil was given to 18 subjects acutely sedated with midazolam following 14 consecutive days of pretreatment with diazepam or triazolam.

In a recent review of the published literature on the use of flumazenil in the treatment of hundreds of patients with drug overdose [20], no significant untoward effects were reported in chronic benzodiazepine users treated with flumazenil.

As detailed in Chapter 1, page 14, flumazenil may actually have a role in resetting benzodiazepine receptors in patients chronically using benzodiazepines.

Conclusions

Benzodiazepines are extensively used for sedation during endoscopy in many countries. Opioids are frequently added but synergism may produce severe respiratory depression. Decreasing the dose of both drugs is strongly recommended when they are combined.

Flumazenil has already been tested and found efficacious and useful in reversing benzodiazepine sedation after endoscopy. The exact role for routine, selective or only emergency use of flumazenil awaits further research and clinical experience.

References

1. Keeffe EB, O'Connor KW (1990) 1989 A/S/G/E survey of endoscopic sedation and monitoring practices. Gastrointest Endosc 36:S13–S18

2. Daneshmend TK, Bell GD, Logan RFA (1991) Sedation for upper gastrointestinal endoscopy. The results of a nationwide survey. Gut 32:12–15
3. McCloy RF, Pearson RC (1990) Which agent and how to deliver it? Scand J Gastroenterol 25 :(Suppl 179): 7–11
4. Trieger N (1989) Intravenous sedation in dentistry and oral surgery. Int Anesthesiol Clin 27 (2):83–91
5. Bell GD, Antrobus JHL, Lee J, Coady T, Morden A (1990) Bolus or slow injection of midazolam prior to upper gastrointestinal endoscopy? Relative effect on oxygen saturation and prophylactic value of supplement oxygen. Aliment Pharmacol Therap 4:393–401
6. Swain DG, Ellis DJ, Bradby H (1990) Rapid intravenous low-dose diazepam as sedation for upper gastrointestinal endoscopy. Aliment Pharmacol Therap 4:43–48
7. Bell GD, Spickett GP, Reeve PA, Morden A, Logan RFA (1987) Intravenous midazolam for upper gastrointestinal endoscopy: a study of 800 consecutive cases relating dose to age and sex of patient. Br J Clin Pharmacol 23:241–244
8. Langendijk PNJ, Loenen AC van, Scheepstra GL, Terhoeve P, de Lange JJ, Danhof M, Breimer DD. Kinetics and effects of midazolam after rectal administration. Submitted for publication.
9. Spear RM, Yaster M, Berkowitz MB, Maxwell LG, Bender KS, Naclerio R, Manolio TA, Nichols DG (1991) Preinduction of anesthesia in children with rectally administered midazolam. Anesthesiology 74:670–674
10. Saint-Maurice C, Landais A, Delleur MM, Esteve C, MacGee K, Murat I. The use of midazolam in diagnostic and short surgical procedures in children. Acta Anaesthesiol Scand (Suppl 92) 34:39–41
11. Walbergh EJ, Wills RJ, Eckhert J (1991) Plasma concentration of midazolam in children following intranasal administration. Anesthesiology 74:233–235
12. Ricou R, Forster A, Bruckner A, Chastonay P, Gemperle M (1986) Clinical evaluation of a specific benzodiazepine antagonist (Ro 15-1788). Studies in elderly patients after regional anaesthesia under benzodiazepine sedation. Br J Anaesth 58:1005–1011
13. Poswillo D (1990) General Anaesthesia, Sedation and Resuscitation in Dentistry: Report of an expert working party (Poswillo Report)
14. Bartelsman JF, Sars PR, Tytgat GN (1990) Flumazenil used for reversal of midazolam-induced sedation in endoscopy outpatients. Gastrointest Endosc 36 (Suppl 3):S9-S12
15. Birch BR, Anson K, Gelister J, Parker C, Miller RA (1990) The role of midazolam and flumazenil in urology. Acta Anaesthesiol Scand (Suppl 92) 34:25–32
16. Rosario MT, Costa NF (1990) Combination of midazolam and flumazenil in upper gastrointestinal endoscopy, a double blind randomized study. Gastrointest Endosc 36(1):30–33
17. Birkenfeld S, Federico C, Dermanski-Arni Y, Bruck R, Melzer E, Bar-Meir S (1989) Double-blind controlled trial of flumazenil in patients who underwent upper gastrointestinal endoscopy. Gastrointest Endosc 35(6):519–522
18. Whitwam JG (1990) Resedation Acta Anaesthesiol Scand (Suppl 92) 34:70–74
19. Dunton AW, Schwam E, Pitman V, McGrath J, Hendler J and Siegel J (1988) Flumazenil: US clinical pharmacology studies. Eur J Anaesthesiol (Suppl 2) 81–95
20. Weinbrum A, Halpern P, Geller E. The use of flumazenil in the management of acute drug posioning – a review. (unpublished)

Section 2: Procedural Safety

Keynote Lecturer

Professor EMMET KEEFFE MD
Professor of Medicine
Oregon Health Sciences University
Portland
Oregon
USA

American Society for Gastrointestinal Endoscopy – Chairman, Standards of Practice Committee

Workshop Director

Professor DAVID FLEISCHER MD FACP
Professor of Medicine
Chief, Endoscopy
Georgetown University
Washington DC
USA

American Society for Gastrointestinal Endoscopy – Treasurer, Member of Executive Committee and Governing Board
Member of Budget and Financial Planning Committee

Workshop Participants

Dr. G. D. BELL, Ipswich, United Kingdom
Dr. A. BENICHOU, Angoulème, France
Dr. M. FRIED, Lausanne, Switzerland
Professor Dr. C. B. H. LAMERS, Leiden, The Netherlands
Dr. P. MATZEN, Hvidovre, Denmark
Dr. J. MONTERO Vazquez, Madrid, Spain
Dr. G. PARK, Cambridge, United Kingdom
Professor Dr. K.-U. SCHENTKE, Dresden, Germany
Dr. L. R. SUTHERLAND, Calgary, Canada

CHAPTER 3

Endoscopic Procedural Safety

Professor EMMET KEEFFE

Introduction

In spite of the remarkable growth in the number and diversity of gastrointestinal endoscopic procedures performed over the past 30 years, endoscopy has maintained a remarkably good overall safety record. Large retrospective surveys of complication rates of specific endoscopic procedures confirm that endoscopy is safe [1–4]. On the basis of variable complication rates of different types of endoscopy, each individual procedure may have its own "inherent" complication rate. For example, diagnostic procedures are safer than therapeutic procedures as is shown for colonoscopy and endoscopic retrograde cholangiopancreatography (ERCP) in Table 1. Complications may also be considered as procedure-related (e.g. bleeding, perforation) or more generic and most often sedation-related (e.g. miscellaneous cardiopulmonary adverse events).

This analysis of procedural safety is an attempt to dissect out factors that are responsible for the low complication rate of gastrointestinal endoscopy. In a sense, endoscopic procedural safety is an opposite consideration from complications of endoscopy. Study of all the potential determinants of procedural safety could lead to recommendations regarding endoscopic practice

Table 1. Complications of endoscopy*

Procedure	Complication (%)	Death (%)
OGD	0.2	0.01
Colonoscopy		
diagnostic	0.4	0.02
polypectomy	2.0	0.05
ERCP		
diagnostic	3.0	0.2
therapeutic	8.0	0.8

* Average figures from large retrospective reviews (see references 1–4).

Table 2. Potential determinants of procedural safety

What	● procedure is performed
	– diagnostic vs. therapeutic
	● conscious sedation or anaesthesia is used
	– benzodiazepine alone vs. with narcotic
	– bolus vs. titration
	● level of monitoring is employed
	– personnel (gastrointestinal nurse) assisting
	– electronic equipment (oximetry, ECG) used
	– resuscitation equipment available
Who	● undergoes endoscopy
	– young vs. elderly
	– medically fit vs. unfit
	– consultation vs. checklist
	● performs endoscopy
	– training and experience
	– competence
Why	● endoscopy is performed
	– indicated vs. contraindicated
Where	● endoscopy is performed
	– fully staffed hospital endoscopy unit vs. small hospital vs. clinic or office
	– elective (endoscopy unit) vs. emergency (ICU or other hospital ward)
How	● the patient is recovered
	● the outpatient is discharged

that might reduce even further the already low complication rate. Much of this analysis of procedural safety is the listing of self-evident elements which contribute to the performance of safe endoscopy. In most cases, there is not a scientific basis to support the contribution of a specific element to the safe performance of endoscopy. Standardisation in endoscopic practice has resulted from generally uniform training in gastrointestinal endoscopy programmes, information available in standard endoscopy textbooks, and policies formulated in endoscopy practice guidelines published by medical societies, such as the American Society for Gastrointestinal Endoscopy (ASGE). In circumstances where endoscopic practice varies, e.g. drugs used for sedation or type of monitoring practices during sedation, the already low complication rate of endoscopy discourages comparative studies, which of necessity must include prohibitively large numbers of patients to determine if a difference exists. Thus, much of this analysis of procedural safety is based on the consensus opinion of expert endoscopists expressing themselves in textbooks and journal topic reviews or through endoscopic society practice guidelines.

There are a large number of potential determinats of procedural safety, which are posed as a series of questions in Table 2. This table serves as an outline of topics to be discussed. Some of the determinants of procedural safety are fixed, e.g. the type of procedure or the medical condition of the patient, while other elements that determine safety are variable, e.g. who performs the procedure, what conscious sedation or monitoring are used, etc. The goal of this review is to provide an overview of factors that determine endoscopic procedural safety.

Potential determinants of procedural safety

What procedure is performed

Procedural safety is largely dependent upon the specific endoscopic procedure performed. This inference is based on complication rates that vary from one procedure to another, and are particularly dependent upon whether a diagnostic or therapeutic procedure is performed. The average figures in Table 1 may be criticised on the basis of being older data reported in many cases during the developmental phase of endoscopy as well as the possibility of under-reporting based on retrospective methodology [4]. However, it is likely that there is an inherent complication rate for each individual endoscopic procedure, which is apparent when comparing the complication and death rates of oesophagogastroduodenoscopy (OGD) to ERCP or comparing diagnostic colonoscopy with colonoscopy with polypectomy. Continued improvements in all aspects of endoscopic practice, even when these changes are adopted on the basis of good medical judgement rather than supported by comparative scientific studies, may be able to lower this "inherent" complication rate of endoscopy. Collection of contemporary data in a prospective fashion from large centres to determine the current complication and death rates of each endoscopic procedure would be a useful foundation on which to compare and evaluate potentially safer new endoscopic practices. The recent development of computerised databases and the new focus on quality assurance may provide further data on the complication rates of individual procedures and stimulate improvements in the provision of endoscopic services.

What conscious sedation or anaesthesia is used

Drugs used for conscious sedation during gastrointestinal endoscopy are the focus of a major section of this Forum on quality control in endoscopy and thus will be discussed here only in the context of procedural safety. Gastrointestinal endoscopy, particularly OGD, can be performed without conscious sedation, but surveys in the United States of America [5] and United Kingdom [6] indicate that both endoscopists and patients prefer some form of premedication. On the other hand, reports from Iraq and Greece of 2,000 and greater than 3,000 patients, respectively, demonstrate that diagnostic OGD can be performed adequately without the need for sedation in from 94% to 98% patients [7, 8]. Whether or not the performance of endoscopic procedures without sedation will predictably increase the endoscopic procedural safety record remains unproven; in fact, anxiety with tachycardia or a more difficult intubation in an alert patient might even precipitate complications and decrease procedural safety. On the other hand, data demonstrating that more than 50% of the deaths associated with OGD are related to cardiopulmonary problems indirectly suggest that endoscopy without sedation may indeed be safer [9].

The typical premedications used for gastrointestinal endoscopy are intravenous benzodiazepines, such as diazepam or midazolam, either alone or with a narcotic such as meperidine (pethidine) [9, 10]. Although the combination of a benzodiazepine with meperidine (pethidine) can achieve a smoother induction,the incidence of respiratory depression is significantly increased [9, 11] (see Chapter 1, page 10–11). The technique and sequence of administration of drugs used for conscious sedation are important additional factors related to the incidence of adverse cardiopulmonary events (see Chapter 2, page 24–25). For example, bolus administration of midazolam is associated with greater oxygen desaturation than slow titrated injection [12]. The ideal rate of titration that is practical and minimises hypoxaemia is not standardised in endoscopic practice, and titration as well as sedation end point should vary according to the pharmacokinetics of the individual drug used for sedation (see Chapter 2, pages 23 and 24). Finally, formal training in techniques of sedating patients is typically not provided during gastrointestinal endoscopy training programmes, but should be considered. In summary, drugs used for conscious sedation play a major role in overall procedural safety, with the use of no drugs possible being the safest and the use of a combination of a benzodiazepine and narcotic being the least safe. The impact of bolus versus titration administration of drugs on procedural safety is unknown.

A number of studies have addressed the physiological changes that occur during gastrointestinal endoscopy, and these data were recently reviewed by Fleischer [11]. The conclusions from the analysis of these studies are as follows:
1. oxygen saturation decreases during OGD and colonoscopy;
2. no correlation exists between the fall in oxygen saturation and the type or dose of medication used and the age or sex of the patient;
3. data are conflicting regarding oxygen saturation and the importance of the length of the procedure, whether sedation is used or not, and if pre-procedural pulmonary function tests are useful; and
4. fall in oxygen saturation is related to the use of larger diameter endoscopes for OGD, the added effect of a narcotic with a benzodiazepine, and the presence of pre-existing pulmonary disease.

The role of using supplemental oxygen in procedural safety remains uncertain. It is clear that significant hypoxaemia during upper and lower gastrointestinal endoscopy can be prevented by pre-oxygenation and/or supplemental nasal oxygen during the procedure [13, 14]. Some endoscopists recommend that prevention of hypoxaemia by use of supplemental oxygen in all high-risk patients is preferable to correction of hypoxaemia after it occurs [13].

It remains to be proven whether or not these different strategies will impact the rate of cardiopulmonary complications associated with endoscopy.

What level of monitoring is employed

Standard clinical monitoring of all patients undergoing gastrointestinal endoscopy has included clinical assessment of tolerance to the procedure and the determination of vital signs at several different junctures before, during and after the procedure by a gastrointestinal nurse. The recent availability of electronic equipment, such as pulse oximetry and continuous electrocardiographic (ECG) monitoring, has led to discussion and debate regarding the usefulness of these procedures and their impact on procedural safety [11]. Additional factors that fueled interest in electronic monitoring were the introduction of midazolam with reports of excessive cardiopulmonary complications during its early use and medico-legal issues. The ASGE responded to the concern of its membership regarding appropriate monitoring policies and developed a guideline entitled "Monitoring of Patients Undergoing Gastrointestinal Endoscopic Procedures" [15]. The key elements of this guideline are as follows:

1. use of special monitoring equipment may be a useful adjunct to patient surveillance, but is never a substitute for clinical assessment;
2. standard clinical monitoring should include the determination of heart rate, blood pressure and respiratory rate before sedation, immediately after the procedure and at the time of discharge from the gastrointestinal unit;
3. the proper role of pulse oximetry and continuous ECG monitoring may be beneficial for certain high risk patients and/or procedures.

The ASGE also sponsored a membership survey of endoscopic sedation and monitoring practices by mail during March and April of 1985 [5]. This survey confirmed the well known fact that most American endoscopists routinely use sedation for endoscopic procedures, excluding flexible sigmoidoscopy. Drugs used regularly included meperidine (pethidine) (87%), midazolam (73%), diazepam (49%) and naloxone (30%). When choosing a benzodiazepine, midazolam was used approximately twice as often as diazepam. The majority of endoscopists had intravenous access available during endoscopy. Premedications for endoscopy were administered by physician endoscopists (83%) or by a registered nurse (43%). Only 3% stated that premedications were given by a physician anaesthesiologist.

This survey indicated that clinical monitoring in the USA during endoscopy was fairly standard with vital signs being monitored before and after the procedure. One gastrointestinal nurse was present during routine procedures, and often a second nurse was present during ERCP. At the time of the survey, 65% of endoscopists used pulse oximetry and 55% used continuous ECG monitoring. However, most electronic monitoring was done in hospitals (99.5%) rather than in private offices (27%). Although the majority of endoscopists use electronic monitoring selectively for high risk patients, 42% used it universally for all patients.

The survey also queried the availability of resuscitation equipment and the qualifications of endoscopists and gastrointestinal nurses in basic life support and advanced cardiac life support. Results indicated that 94% of endoscopists had resuscitation equipment ("crash cart") in the endoscopy unit, and 98% had one available in the area or building. A surprising finding from the survey was that only 76% of endoscopists were certified in basic life support and 30% in advanced life support techniques.

The widespread availability of electronic monitoring equipment and its likelihood to enhance the safety of conscious sedation, as well as a number of pressures such as medico-legal considerations, is leading to the use of electronic monitoring on a routine rather than selective basis.

Who undergoes endsocopy

Endoscopic procedural safety is certainly also dependent upon the individual patient undergoing endoscopy. Although there are no specific data addressing this issue, it is obvious that the young and medically fit patient is likely to tolerate an endoscopic procedure better than the elderly or medically unstable patient. Endoscopists typically assess risk based on global assessment during pre-procedure evaluation. The risk of endoscopy, however, might be better defined by routine use of standardised risk classifications, such as those used by anaesthesiologist (ASA physical status, see Chapter 4, Table 1) or by intensive care medicine physicians (APACHE II score) [16, 17].

The role of familiarity with the patient prior to endoscopy in procedural safety is uncertain. Endoscopic procedures may be performed on patients first seen for complete consultation prior to endoscopy. Other patients may undergo endoscopy without consultation; in these patients, some relevant medical history related to the risk of endoscopy is usually obtained. Typically the endoscopist and/or gastrointestinal nurse completes a checklist containing elements such as current medications, allergies, bleeding disorders, infections (hepatitis, HIV), and heart or lung disease. Whether or not obtaining prior consultation versus simply doing a pre-endoscopy medical checklist contributes to the performance of safe endoscopy remains uncertain.

The education and preparation of patients before endoscopy may also be factors that are important in the completion of a safe procedure. Certainly education allays anxiety and may enhance patient cooperation and the likelihood of an uncomplicated procedure. Ensuring an empty stomach or a clean colon facilitates a smoother, and probably safer, endoscopy or colonoscopy. Instructing patients to avoid aspirin or nonsteroidal anti-inflammatory drugs may decrease the incidence of bleeding after tissue biopsy or therapeutic procedures such as polypectomy or sphincterotomy. Identification of patients who warrant antibiotic prophylaxis may be important. Although there are no firm data that have clearly established the benefit of using prophylactic antibiotics, adherence to standard recommendations for the prevention of bacterial endocarditis seems appropriate [18]. Finally, special preparation in certain circumstances, such as airway protection for active upper gastrointestinal haemorrhage, may prevent specific complications such as aspiration pneumonia.

Who performs endoscopy

Endoscopic training and experience are major determinants of subsequent competence in practice. The ASGE has published a guideline entitled "Standard of Practice of Gastrointestinal Endoscopy" which outlines that physicians and surgeons who practise gastrointestinal endoscopy should meet a number of standards related to training, practice, hospital privileges, continuing education, and performance review [19]. The specific methods of granting hospital privileges to perform gastrointestinal endoscopy, including minimum numbers of procedures performed in training, has also been published by the ASGE [20]. The American College of Physicians has also developed a series of guidelines to assist in the assessment of physician competence on a procedure-specific basis [21–24].

The Standards of Training Committee of the ASGE has analysed the training process in depth and recently prepared a new document entitled "Principles of Training in GI Endoscopy" [25]. This document defines how endoscopic skills are gradually acquired under appropriate supervision, which is initially complete and later periodic. Training in standard gastrointestinal endoscopic procedures is distinguished from advanced training in procedures such as ERCP, laparoscopy, dilatation for achalasia, oesophageal stent placement and endoscopic tumour ablation. It may be appropriate that only selected trainees undergo advanced endoscopic training and that they extend their period of training for one or two additional years. This policy for training might slow the pace of an emerging situation in metropolitan areas in the USA where an increased number of practising endoscopists are performing fewer advanced procedures.

The most controversial aspect of the ASGE document on training is the definition of a threshold number of procedures for assessing competence (Table 3). Although the number of procedures which a trainee must perform to attain competence is variable, threshold numbers are offered for endoscopic training directors to subsequently assess and certify competence. Completion of a minimum number of procedures does not by itself imply competence.

Endoscopic procedural safety is likely to be enhanced by the appropriate training of physicians or surgeons in the cognitive and technical aspects of endoscopy. The ASGE has always supported the position that individuals

Table 3. Threshold for assessing competence

	Number of cases
Standard Procedures	
Diagnostic OGD	100
Total colonoscopy	100
Snare polypectomy	20
Nonvariceal haemostasis (upper and lower; includes 10 active bleeders)	20[1]
Variceal haemostasis (includes 5 active bleeders)	15
Oesophageal dilatation with guide wire	15
Flexible sigmoidoscopy	25
Percutaneous endoscopic gastrostomy	10
Advanced Procedures	
ERCP (diagnostic)	75
ERCP (therapeutic)	25[2]
Tumour ablation	20
Pneumatic dilatation for achalasia	5
Laparoscopy	25
Oesophageal stent placement	10

[1] Included in total number
[2] Includes 20 sphincterotomies and 5 stent placements and is in addition to the 75 diagnostic ERCP procedures.

performing gastrointestinal endoscopy should be trained in endoscopy as part of a broader clinical discipline such as gastroenterology or gastrointestinal surgery. Comprehensive training in endoscopy can no longer be acquired outside of an accredited training programme; in particular, endoscopic hospital privileges should not be granted to applicants citing only attendance in short courses as the sole training experience [26]. Moreover, hospital privileges should be granted for each separate procedure for which training and competence has been documented rather than globally for all endoscopic procedures. Finally, periodic review of endoscopic hospital privileges is appropriate to review continuing performance.

An important technique to assess ongoing competence in gastrointestinal endoscopy is quality assurance. According to the Joint Commission on Accreditation of Healthcare Organizations (JCAHO), a quality assurance programme should be "designed to objectively and systematically monitor and evaluate the quality and appropriateness of patient care, pursure opportunities to improve patient care, and resolve identified problems" [27]. The ASGE, in consultation with the JCAHO, developed and published a document entitled "Quality Assurance of Gastrointestinal Endoscopy" [28]. Quality assurance is now widely used in the USA for hospital recredentialing for continued performance of individual endoscopic procedures and for the identification and correction of deficiencies and problems. The major elements of a quality assurance programe in gastrointestinal endoscopy are:

1. the procedure report;
2. the endoscopic unit record; and
3. procedure review.

The endoscopy unit record serves as the database for the procedure review process. A number of clinical indicators such as the appropriateness of an endoscopic procedure, absence of contraindications, technical performance, or complications are items that may be reviewed periodically.

It remains to be defined what minimum number of procedures should be performed annually to maintain skills for each type of endoscopy. Committees of several societies are beginning to address this issue and develop a consensus opinion regarding maintenance of competence. For example, performance of fewer than 10 endoscopic sphincterotomies per year is probably not adequate to maintain proficiency and procedural safety. To use an analogy, I find that I can ride my bicycle very competently on the first day of spring after a long winter break, but that I need considerable practice to reach the previous season's level of competence in tennis. Thus, endoscopists require variable ongoing procedural experience from very little to considerable to maintain competence.

A final subjective consideration that is strongly related to the complication rate or safety of endoscopy is clinician judgment regarding how aggressively an individual procedure is performed. The clinical situation, including the alternative diagnostic and therapeutic approaches and risk versus benefit, must always be weighed against continuation or completion of a difficult procedure, e.g. reaching the caecum during colonoscopy or cannulating the bile duct during ERCP.

In summary, initial training and competence in gastrointestinal endoscopy, ongoing review of performance, long-term maintenance of competence and clinician judgement are factors that are criticially important determinants of procedural safety.

Why endoscopy is performed

Procedural safety may be determined in some part by knowledge and adherence to the indications and contraindications for individual procedures [29]. Although some data suggest that at least OGD may be performed inappropriately in up to 17% of cases [30], this finding, even if true, may not directly relate to procedural safety. However, careful attention to the indications for endoscopy and, more importantly, avoiding procedures that are contraindicated should result in safer endoscopy. Upper gastrointestinal endoscopy is generally contraindicated when:
1. the risk of the procedure outweighs the potential benefit;
2. the patient is unable to cooperate; and
3. a perforated viscus is known or suspected [29].

In addition, lower endoscopy is contraindicated in fulminant colitis and acute severe diverticulitis [29].

Where endoscopy is performed

Procedural safety is likely to be determined by the setting in which endoscopy is performed, i.e. a fully staffed hospital endoscopy unit versus a small rural hospital versus an outpatient clinic or office. In addition, elective endoscopic procedures performed as an emergency in an intensive care unit, emergency ward or other hospital location. The ASGE has recently revised guidelines for the establishment of gastrointestinal endoscopy areas [31]. Safe and efficient performance of gastrointestinal endoscopy depends upon many of the issues under discussion, including:
1. properly trained and competent endoscopists;
2. properly trained and competent gastrointestinal nurses and supportive ancillary personnel;
3. functioning and well maintained equipment;
4. adequately furnished preparation, endoscopy and recovery areas;
5. equipment and trained personnel to perform cardiopulmonary resuscitation; and
6. a functioning quality assurance programme [28].

The lighting in the endoscopy room may also be an important safety factor. A light or dim room allows better clinical monitoring of patient status than a darkened endoscopy area.

There is a trend in the USA and other countries to perform endoscopy in out-of-hospital endoscopy areas. The ASGE guideline on the establishment of gastrointestinal endoscopy areas states that "standards for out-of-hospital endoscopic practice should be identical to the recognized guidelines followed in the hospital" [31]. The recent ASGE survey of endoscopic sedation and monitoring practices, however, reveals that usage of electronic monitoring devices was much higher for hospital procedures (99.5%) as opposed to office procedures (27%) [5]. There are no published data comparing complications, in particular cardiopulmonary complications, in various hospital settings versus out-of-hospital settings, but differences in monitoring by trained personnel and use of electronic equipment may be associated with differences in the safety record in the various areas where endoscopy is performed.

The important contribution of a trained gastrointestinal endoscopy nurse in endoscopic procedural safety cannot be overemphasised. Since the endoscopist is usually focused on the technical aspects of the procedure, the primary role for patient monitoring rests with the gastrointestinal nurse. An issue of potential relevance to procedural safety is the presence of two nurses in each endoscopy room when therapeutic procedures are performed. One nurse should be assigned primary responsibility for monitoring the patient, while the other nurse assists the endoscopist in the retrieval of equipment necessary for the completion of successful therapeutic procedures. Adequate staffing by trained nurses is particularly important after hours when emergency and therapeutic procedures are often performed. Many hospitals, particularly teaching hospitals, have well staffed endoscopy units during regular hours but make use of trainees and/or untrained ward nurses to assist with complex therapeutic procedures after hours.

An ancillary issue related to where endoscopy is performed is the appropriate cleaning, sterilisation and disinfection of endoscopic instruments [32]. Regular adherence to strict protocols for the mechanical cleaning and chemical disinfection of instruments is important to reduce or eliminate the likelihood of transmission of infection to patients undergoing endoscopic procedures.

How the patient is recovered and discharged

The recovery and discharge policy for inpatients and/or outpatients undergoing endoscopic procedures is often not standardised. The monitoring of patients recovering from endoscopy and conscious sedation is often dependent upon the availability of space and personnel. In some hospitals, the recovery process takes place in a separate, short stay or recovery unit analogous to the recovery of patients following general anaesthesia. In other hospitals, recovery takes place in an ancillary unit with variable monitoring. The risk of conscious sedation certainly extends beyond the time of the endoscopic procedure, and monitoring should persist until the patient has stable vital signs and is awake and alert.

Generally patients are discharged when monitoring is no longer necessary by the above criteria, and they have met the additional requirement of ambulation. Discharge policies for patients undergoing outpatient endoscopy should be written and standardised. Patients having conscious sedation should be accompanied by a friend or relative and instructed not to drive, operate machinery or drink alcohol for 24 hours.

Summary

In conclusion, there are a large number of potential determinants of endoscopic procedural safety. Standard procedures in most endoscopy units contain a number of important elements that are likely to contribute to the overall low complication rate of endoscopy. Strict adherence to the recommendations found in endoscopic practice guidelines regarding conscious sedation, monitoring, endoscopic training and competence, indications and contraindications, endoscopic unit staffing, and recovery and discharge policies may result in a reduction in the "inherent" complication rate of individual endoscopic procedures. Ongoing review of endoscopic practice by hospital quality assurance programmes may result in continued improvement in the overall safety record. Endoscopic procedural safety has not received appropriate attention in the published literature. The many questions raised in this summary will hopefully stimulate prospective trials to provide a better scientific basis for current standards of practice related to safety and possibly improve yet further the already good record.

References

1. Carey WD (1987) Indications, contraindications, and complications of upper gastrointestinal endoscopy. In: Sivak MV Jr., (ed.) Gastroenterologic Endoscopy. Philadelphia: WB Saunders: 296–306
2. Ferguson DR, Sivak MV Jr (1987) Indications, contraindications, and complications of ERCP. In: Sivak MV Jr., (ed.) Gastroenterologic Endoscopy. Philadelphia: WB Saunders: 581–598
3. Rankin GB (1987) Indications, contraindications and complications of colonoscopy. In: Sivak MV Jr., (ed.) Gastroenterologic Endoscopy. Philadelphia: WB Saunders: 868–880
4. Schrock TR (1989) Complications of gastrointestinal endoscopy. In: Sleisenger MH, Fordtran JS, (eds.) Gastrointestinal Disease. 4th ed. Philadelphia: WB Saunders: 216–222
5. Keeffe EB, O'Connor KW (1990) 1989 A/S/G/E survey of endoscopic sedation and monitoring practices. Gastrointest Endosc 36:S13–S18
6. Daneshmend TK, Bell GD, Logan RFA (1991) Sedation for upper gastrointestinal endoscopy: results of a nationwide survey. Gut 32:12–15
7. Al-Atrakchi HA (1989) Upper gastrointestinal endoscopy without sedation: a prospective study of 2000 examinations. Gastrointest Endosc 35:78–81
8. Ladas SD, Giorgiotis C, Pipis P, et al (1990) Sedation for upper gastrointestinal endoscopy: time for reappraisal? Gastrointest Endosc 36:417–418
9. Bell GD (1990) Review article: premedication and intravenous sedation for upper gastrointestinal endoscopy. Aliment Pharmacol Therap 4:103–122
10. Ross WA (1989) Premedication for upper gastrointestinal endoscopy. Gastrointest Endosc 35:120–126
11. Fleischer D (1989) Monitoring the patient receiving conscious sedation for gastrointestinal endoscopy: issues and guidelines. Gastriontest Endosc 35:262–266
12. Bell GD, Antrobus JHL, Lee J, Coady T, Morden A (1990) Bolus or slow titrated injection of midazolam prior to upper gastrointestinal endoscopy? Relative effect on oxygen saturation and prophylactic value of supplemental oxygen. Aliment Pharmacol Therap 4:393–401
13. Bell GD, Brown NS, Morden A, Coady T, Logan RFA (1987) Prevention of hypoxaemia during upper gastrointestinal endoscopy by means of oxygen via nasal cannulae. Lancet 1:1022–1024
14. Gross JB, Long WB (1990) Nasal oxygen alleviates hypoxemia in colonoscopy patients sedated with midazolam and meperidine. Gastrointest Endosc 36:26–29
15. Standards of Practice Committee. American Society for Gastrointestinal Endoscopy. Monitoring of patients undergoing gastrointestinal endoscopic procedures. Guidelines for clinical application. Gastrointest Endosc 37:120–121
16. Schneider AJL (1983) Assessment of risk factors and surgical outcome. Surg Clin N Amer 63:1113–1116
17. Knaus WA, Draper EA, Wagner DP, Zimmerman JE (1985) APACHE II: severity of disease classification system. Crit Care Med 13:818–829
18. Dajani AS, Bisno AL, Chung KJ, et al (1990) Prevention of bacterial endocarditis. Recommendations by the American Heart Association. JAMA 264:2919–2922

19. Standards of Training and Practice Committee. American Society for Gastrointestinal Endoscopy. Standards of practice of gastrointestinal endoscopy. Guidelines. Gastrointest Endosc 34:8S
20. Standards of Training and Practice Committee (1988). American Society for Gastrointestinal Endoscopy. Methods of granting hospital privileges to perform gastrointestinal endoscopy. Gastrointest Endosc 34:28S–29S
21. Health and Public Policy Committee (1987). American College of Physicians. Clinical competence in the use of flexible sigmoidoscopy for screening purposes. Ann Intern Med 107:589–591
22. Health and Public Policy Committee (1987). American College of Physicians. Clinical competence in colonoscopy. Ann Intern Med 107:772–774
23. Health and Public Policy Committee (1987). American College of Physicians. Clinical competence in diagnostic esophagogastroduodenoscopy. Ann Intern Med 107:937–939
24. Health and Public Policy Committee (1988) American College of Physicians. Clinical competence in diagnostic endoscopic retrograde cholangiopancreatography. Ann Intern Med 108:142–144
25. Standards of Training Committee (1991). American Society for Gastrointestinal Endoscopy. Principles of training in GI endoscopy. Manchester, Massachusetts: American Society for Gastrointestinal Endoscopy
26. Standards of Training and Practice Committee (1988). American Society for Gastrointestinal Endoscopy. Statement on role of short courses in endoscopic training. Gastrointest Endosc 34:14S–15S
27. Joint Commission on Accreditation of Healthcare Organizations. Accreditation manual for hospitals, 1990 (1989). Chicago: Joint Commission on Accreditation of Healthcare Organizations
28. American Society for Gastrointestinal Endoscopy. Quality assurance of gastrointestinal endoscopy. Manchester, Massachusetts: American Society for Gastrointestinal Endoscopy
29. Committee on Endoscopic Utilization and Standards of Practice Committee. American Society for Gastrointestinal Endoscopy (1989). Appropriate use of gastrointestinal endoscopy. Manchester, Massachusetts: American Society for Gastrointestinal Endoscopy, May
30. Kahn KL, Kosecoff J, Chassin MR, Solomon DH, Brook RH (1988) The use and misuse of upper gastrointestinal endoscopy. Ann Intern Med 109:664–670
31. Standards of Practice Committee. American Society for Gastrointestinal Endoscopy (1989). Guidelines for establishment of gastrointestinal areas. Manchester, Massachusetts: American Society for Gastrointestinal Endoscopy, August
32. Standards of Training and Practice Committee. American Society for Gastrointestinal Endoscopy (1988). Infection control during gastrointestinal endoscopy. Guidelines for clinical application. Gastrointest Endosc 34:37S–40S

CHAPTER 4

Report of Workshop on Procedural Safety

Professor DAVID FLEISCHER

Pre-procedural safety

Clinical competence and judgement are issues which must be resolved by adequate training before the procedure (see Chapter 3, pages 39–41). However, there are a variety of practices which can be undertaken before an endoscopy that may influence the safety of that procedure.

Clinical evaluation

It is appropriate for a medical history to be taken prior to an endoscopy. The Workshop participants felt that it was important for there to be a standardised record kept of that history. It is extremely helpful if there is a check list that must be completed. The endoscopist needs to know, for example, about current medication, drug allergy and bleeding diathesis. However, is it necessary for a physical examination to be performed? To the anaesthesiologist in the Workshop, it was an anathema that someone should consider doing a procedure without listening to the patient's heart or lungs. In fact, the gastroenterologists felt that on most occasions an examination had already been performed by another clinician and the number of times that examination influenced the management of the procedure was very small. The Workshop participants were unanimous that there was no indication for any routine blood tests.

Risk assessment

Is it possible to assign a patient to a risk category before the procedure to allow the endoscopist to modify his procedural practice if necessary? Anaesthesiologists have a standard – the American Society of Anesthesiology (ASA) classification [1] (Table 1). The APACHE II score [2] is another way of assessing illness. This is based on a complex scoring and calculation system developed for patients on intensive care units. The system involves scoring 12 acute physiological variables, age points and chronic health points. The risk of death in a particular patient can be computed by weighting the total score with a diagnostic category and the patient's status, i.e.

Table 1. American Society of Anesthesiologists (ASA) Physical Status

I	Healthy patient	
II	Mild systemic disease	absent or only slight functional limitation
III	Severe systemic disease	definite functional limitation
IV	Severe systemic disease	constant threat to life
V	Moribund patient	unlikely to survive 24 hours with or without operation

Schneider AJL. Surg Clin N Amer 1983; 63: 1113

non-operative or post-operative. The Workshop members were uncertain as to the relevance of these classifications to endoscopy. Dr. Matzen from Denmark was the only gastroenterologist in the Workshop who assesses risk factors formally using a specially designed scoring system (Table 2). Patients are categorised as low, intermediate or high risk. Points are allocated for each of these categories according to the patient's condition and the procedures being considered. If someone is having a colonoscopy and they have some ongoing illness that is compensated and fairly stable that would be a score of $2 + 2 = 4$. The degree of monitoring that would be used varies from simple observation to having an anaesthesiologist present. In this example, the patient would require extra assistants and supplemental oxygen.

Table 2. Pre-endoscopy evaluation risk category: add patient and procedure score

Risk	Patient	Procedure
Low (score 1)	No impairment of vital organs	Upper endoscopy Sigmoidoscopy
Intermediate (score 2)	Compensated conditions	ERCP Colonoscopy
High (score 3)	Decompensated conditions	Therapeutic endoscopy

Recommended level of monitoring according to risk category:
Added score 2–3 = Simple observation
Added score 4–5 = 2 assistants, oxygen
Added score 6   = Anaesthesiology level

Developed by Matzen

The Workshop members recommended that this approach is logical and advisable, particularly if the endoscopist is to use selective rather than universal decision making whether or not to use monitors, or what medication to give. We were persuaded and we learned from our anaesthesiology colleagues.

Once a procedure has been scheduled, the endoscopist feels obliged to proceed. If the patient's condition changes or the procedure is no longer indicated, the endoscopist needs to have the courage to cancel the procedure, even if, for example, the patient has taken a purgative solution in preparation for colonoscopy.

Life support

In different countries there are different degrees of liberty with which non-anaesthesiologists are allowed to give anaesthetic drugs or practise conscious sedation, but, irrespective of the circumstances, there is no question that the equipment to perform cardiopulmonary resuscitation must always be available, regularly checked and in working order. There are two separate training levels for life support.

In the United States there is basic life support training which entails learning how to do mouth to mouth resuscitation, chest compression and maintaining the airways – the standard ABC of basic life support. There is also advanced training that deals with reading electrocardiograms, intubating patients, and providing intravenous access. In Georgetown Hospital, (Washington, D.C., USA), it is a requirement that a clinician who practises endoscopy must not only be trained in basic life support but also in advanced cardiac life support. The Workshop participants concluded that all endoscopists and their assistants should be at least trained in basic life support. This requirement would enhance patient safety.

Safety during the procedure

Intravenous access

Ninety three percent of clinicians in the United States currently have some intravenous access during an endoscopic procedure. Whilst practices vary among countries, the Workshop recommended that continuous intravenous access should be used when sedation is employed. Some endoscopists, who do regular diagnostic endoscopies which may take only five or ten minutes, feel it is not necessary to have continuous intravenous access when administering a single injection of a sedative drug. The Workshop members did not support this practice since, whenever there is an opportunity for a serious adverse event, the endoscopist must be in a position to deal with this without delay.

Monitoring

Should all patients, irrespective of their risk group, be monitored to the same degree or should there be selective monitoring? As discussed above, one approach is that of Matzen to define high risk patients who require extensive monitoring. However, it may be that this risk category assessment may not identify the patients who are going to get into trouble. Therefore, the alternative approach is to use universal monitoring. Within the Workshop, four out of the ten gastroenterologists used universal and six were selective. This difference of opinion is probably quite appropriate since there are no data to show that one approach is better than the other.

Interpretation of the results of monitoring

Using pulse oximetry, it is possible to develop an algorithm for use in the endoscopy unit to identify that a certain oxygen saturation (maybe SaO_2 of 90% for example) should be attended to by either deep breathing or by supplemental oxygen if it is not being used. However, electrocardiographic monitoring and blood pressure are open to a wider interpretation. With continuous ECG monitoring a number of more minor abnormalities can be identified, e.g. tachycardia and ectopic beats, but are they important and do they require a response? How great a fall of blood pressure is of concern given the patient's underlying medical condition and when should a therapeutic response be made? Thus it is harder to develop an algorithm approach for ECG and blood pressure than it is for pulse oximetry.

Alternatively, the identification of trends in readings from monitoring is likely to be more important than standardised threshold levels which only identify the situation at a single moment in time. On the basis of trends, it is easier to interpret the data.

Supplemental oxygen

Current endoscopic practice generally favours the selective use of supplemental oxygen. However, there is a theoretical concern that carbon dioxide retention could be masked. On the other hand, supplying oxygen to a patient is so cheap, relative to the procedure, and the potential benefits of shifting the oxygen/haemoglobin dissociation curve to the left, argue in favour of universal use of supplemental oxygen.

Nursing support

The majority of the Workshop members felt that one assistant was required in addition to the endoscopist for diagnostic endoscopy and colonoscopy and almost everyone used two for the more complicated procedures. The British Society of Gastroenterology recommends that two assistants be present for all procedures [3].

Risk benefits

Endoscopists should be aware that the assessment of risks versus benefits should be continuous from the pre-procedure stage and throughout the procedure. If after one hour into a colonoscopy the endoscopist has still not achieved his end point, he should ask himself if it is his ego that is driving him, or is there a really good medical indication for continuing. If endoscopists undertook this continuous reassessment then procedures would be safer.

Post-procedural safety

Use of flumazenil

The majority of endoscopists who have flumazenil available in their countries, reverse benzodiazepine-induced sedation selectively. Flumazenil could be used for routine reversal of sedation in order to improve throughput and reduce requirements for recovery (see Chapter 1, page 12). All the members of the Workshop who currently use flumazenil felt that cost was inhibiting their use of the drug, and that further economic appraisals are necessary.

Post-procedural monitoring

If you monitor a patient during a procedure do you monitor them in the recovery room? The reason this is important is because a lot of the complications associated with intravenous sedation have occurred, not during the procedure itself, but during recovery. It is interesting that although all the Workshop members monitored patients during the procedure, only half of the group monitored patients during recovery. Within the Workshop certain end points as to when the monitoring should cease were defined. These were when the patient had stable vital signs, could speak clearly, understand and sit up.

Discharge instructions

The fact that the patient who has undergone conscious sedation can leave the endoscopy recovery area when these recovery end points have been reached and can walk around is no excuse for not giving written discharge instructions. Furthermore, the patient should not drive home, make important decisions nor drink alcohol. The length of time that these precautions need to be observed remains a matter of debate in the absence of adequate published data, and also relates to the length of action of the benzodiazepine used and its metabolites.

Quality assurance

The American Society for Gastrointestinal Endoscopy has published guidelines [4], endorsed by other societies, which address specific areas of documentation which will improve the quality assurance of endoscopic procedures. A legible report must be written after every procedure. There should also be an up-to-date log book available to anyone who wishes to refer to previous cases. The Workshop members agreed that the rigorous compilation of this log book is likely to increase patient safety by influencing the way that clinicians perform procedures. A monthly conference should take place at which complications are discussed and quality assurance issues are addressed. At such a conference in Georgetown University (Washington, D.C.), all complications from the preceding month are described and discussed in a constructive fashion to ascertain if and how they could have been avoided.

The future

The Workshop participants identified many areas where definitive data are lacking and recommended that studies be undertaken to provide the answers.
1. Prospective studies to address the complication rates for endoscopic procedures – identify the complications, how many are cardiorespiratory, where they occur, how could they have been avoided, and which sedative drugs were administered.
2. Each country should conduct a survey like Keeffe's survey of the endoscopic practices in the United States.
3. Studies should be undertaken to provide outcome data which is more specific than simply looking at mortality, for example, does neurological impairment occur after the procedure?
4. Prospective studies measuring the incidence of cardiac arrythmias and cerebral function with and without oximetry. Does oximetry help us at all or does it really not provide much benefit?
5. By using supplemental oxygen prior to a procedure, are we helping our patients or are we simply making ourselves more comfortable?
6. The development of a standardised risk assessment index in order to resolve the dilemma of selective versus universal monitoring.
7. Studies on the post-procedural use of flumazenil to assess its effects on oxygen saturation and to evaluate its cost benefit.

References

1. Schneider AJL (1983) Assessment of risk factors and surgical outcomes. Surg Clin N Amer 63:1113
2. Knaus WA, Draper EA, Wagner DP, Zimmerman JE (1985) APACHE II: a severity of disease classification system. Crit Care Med 13:818–829
3. Bell GD, McCloy RF, Charlton JE, Campbell D, Dent NA, Gear MWL, Logan RFA, Swan CHJ (1991) British Society of Gastroenterology. Recommendations for standards of sedation and patient monitoring during gastrointestinal endoscopy. Gut 32:823–827
4. Standards of Practice Committee American Society for Gastrointestinal Endoscopy (1991). Monitoring of patients undergoing gastrointestinal endoscopic procedures. Guidelines for clinical application. Gastrointest Endosc 37:120–121

Section 3: Resource Management

Keynote Lecturer and Workshop Director

Professor ALAN MAYNARD BA BPhil
Professor of Economics
Centre for Health Economics
University of York
York
United Kingdom

Director of The Department of Health-Economic and Social Research Council Centre for Health Economics at the University of York. The Centre for Health Economics is a World Health Organisation Collaborating Centre in Psychosocial and Economic Aspects of Health.

Workshop Participants

Dr. R. DE PEYER, Geneva, Switzerland
Dr. F. HABAL, Toronto, Canada
Dr. G. D. KERR, Shrewsbury, United Kingdom
Dr. A. KRUSE, Aarhus, Denmark
Dr. R. MILLER, London, United Kingdom
Professor T. G. PARKS, Belfast, Northern Ireland
Dr. P. SLEZAK, Stockholm, Sweden
Dr. T. VALLOT, Paris, France

CHAPTER 5

Economic Aspects of Endoscopy

Professor ALAN MAYNARD

Introduction

Endoscopy, like many other activities in health care, is rarely evaluated from an economic perspective. The objective of economic analysis is to facilitate choice by identifying the value of what is given up when a procedure is provided (the opportunity cost) and what is gained, ultimately in enhancements in the length and quality of life (the outcome).

In a world of scarcity it is essential that decision makers in the health care system behave efficiently. This requires them to use only those procedures which give maximum outcome at least cost. If practitioners do not behave efficiently they deprive potential patients (e.g. on the waiting list) of care from which they could benefit. Such behaviour is unethical as well as being inefficient.

The process of identifying efficient practices in any health care system is difficult because of the absence of relevant data. Most decision making in all health care systems takes place in a "data free" environment where "guestimates" and values overwhelm logic and careful measurement. To change this behaviour, careful collaborative research is needed.

A review of the paucity of relevant data to inform medical choices is presented, followed by an outline of the economic approach for the evaluation of medical practice.

Health care choices are ill-informed

What is the problem?

Any attempt to identify the costs and outcomes of medical practices is inhibited by the absence of relevant data about inputs, processes and activities, and outcomes. These data need to be collected so that the relationships between them can be explored (Fig. 1).

Costs data

On the inputs side costs have to be identified, measured and valued. The costs of an endoscopy procedure may accrue to a variety of groups and the

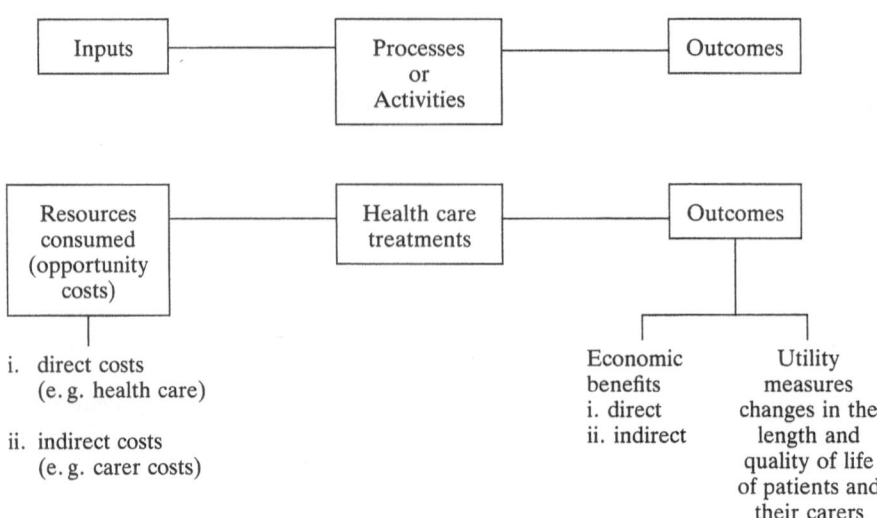

Fig. 1. The framework for an economic evaluation

process of identifying these costs may be complex. Obviously there are costs for the hospital: the value of doctors' time, nurses' time, drugs, equipment, heating, lighting and cleaning, and the capital cost of the building. There are also costs to the patients and their relations or friends: the value of the time taken to get to the hospital, undergo the procedure and recover and anyone who accompanies them. How quickly do they recover and return to everyday living, including work?

Whilst these costs can be identified and measured quite easily in principle, it is commonplace for routine health care data systems not to have this information. As a consequence any study of endoscopy requires a special attempt to identify and measure the relevant costs.

Valuing these costs is not always easy. The valuation of some costs can be carried out according to agreed conventions about how a "good study" is to be executed. For instance, the value of the doctor's time is measured by using the hourly wage rate; economists asserting that wages reflect the value of the marginal product of labour. To the extent that wages are not good measures of the value of the marginal product of labour, this method gives biased results.

There are no agreed best practices for valuing leisure time. If a wife accompanies her husband to endoscopy and she gives up leisure time to do this, what is the opportunity cost of her caring for her spouse? Clearly there is a cost which can be identified and measured, she might otherwise be shopping or doing the ironing, but what is its value?

Health care managers rarely collect the cost data required for economic evaluation. They tend to collect information on financial flows, prices and hospital tariffs, which they need to balance their books: the requirements of

Table 1. Plausible sources of variation at different levels of aggregation

Variation between	Morbidity	Supply	Clinical	Demand
GPs	S	0	L	S
Districts	M	M	L	S
Regions	L	L	S	M
Countries	L	L	L	L

L = Large; M = Medium; S = Small; 0 = No effect relative to others in row
Source: McPherson, K (1990)

financial control do not produce cost data which facilitate economic appraisal.

Furthermore, all cost data are collected in a fragmented way for each component part of the health care system. Data are not linked so that it is not possible to determine the cost of the diagnostic or treatment episode.

Process and activity data

Whilst there is a paucity of relevant input (cost) data there are often better data on activities, e.g. the number of procedures carried out. From these data it is possible to identify enormous variations in medical practice. In surgery and medicine particular groups of patients with the same diagnoses may be treated very differently between practitioners, between hospitals, between regions and between countries.

These variations in practice have been the subject of considerable analysis. For instance, Priestman and his colleagues [1] sent questionnaires to all UK radiotherapists asking them to indicate how they would treat 6 postulated patients with particular cancers at given stages and with given complications. The results showed large variations in treatment and in whether treatment was offered. Similar variations are evident in the surgical treatment of breast cancer where there is no evidence that survival after radical mastectomy, with its severe effects on the (psychological) quality of life of patients, differs from that for lumpectomy for most patients.

McPherson [2] has reviewed the literature on variations in medical practice between general practitioners, districts, regions and countries and sought to attribute their causes to differences in supply, demand, morbidity and clinical practice (Table 1). The importance of variations in clinical practice shows that the basis of much treatment is unproven and a matter of debate and experimentation for many practitioners.

Outcome data

Given the significant variation in diagnostic, treatment and preventive practice it is surprising that more effort is not made to identify, measure and value outcomes. Much medical practice is focused on process and structure: "the operation was a success but the patient died"!

Outcomes, enhancements in the length and quality of life of patients, are not easy to measure. Survival data are poor and, as with cost data, there is little effort to link data to produce survival data for treatment episodes across many years. Hospitals, managers and clinicians may not even collect and evaluate in-patient survival data.

The need for such data has been recognised for some time:

> "All public institutions must be compelled to keep case-books and registers, on a uniform plan. Annual abstracts of the results must be published. The annual medical report of cases must embrace hospitals, lying-in hospitals, dispensaries, lunatic asylums and prisons."
>
> Lancet (1841)

The advocacy of the editor of the Lancet contributed to the formulation of the 1844 Lunacy Act which required all public psychiatric hospitals to collect outcome data in terms of whether patients were dead, relieved or unrelieved. Such data were collected throughout the nineteenth century. Florence Nightingale advocated the use of this classification in 1863, concluding:

> "I am fain to sum up with an urgent appeal for adopting this or some uniform system of publishing the statistical records of hospitals. There is a growing conviction that in all hospitals, even in those which are best conducted, there is a great and unnecessary waste of life ...
>
> In attempting to arrive at the truth, I have applied everywhere for information, but in scarcely an instance have I been able to obtain hospital records fit for any purpose of comparison. If they could be obtained, they would enable us to decide many other questions besides the ones alluded to. They would show subscribers how their money was being spent, what amount of good was really being done with it, or whether the money was doing mischief rather than good."
>
> Florence Nightingale (1863)

Little progress has been made in the last 130 years. Data which indicate the enhancement in the length of life resulting from medical interventions are few. Furthermore, such data alone are inadequate measures of outcome. If only the duration of survival is used as a basis for evaluating outcome and if these data were used to determine resource allocations, investments would be made in life saving interventions at the expense of those interventions which affect the quality of life, i.e. transplants rather than hip replacements would be funded. To avoid this bias in resource allocation it is necessary to measure outcomes both in terms of enhancements in the duration and quality of survival.

But how is the quality of survival to be measured? The measurement of the quality of life involves two steps: identifying and agreeing descriptors, and valuing (ranking) alternative combinations of descriptors. There are not agreed 'gold standards' for descriptors of valuation methods.

The descriptors for quality of life measurement will typically include items which describe physical, social and psychological well-being. The physical

Table 2. The Rosser Descriptors: Disability and Distress

Disability		
I	No disability	A. No Distress
II	Slight social disability	B. Mild
III	Severe social disability and/or slight impairment of performance at work Able to do all housework except very heavy tasks	C. Moderate
IV	Choice of work or performance at work severely limited Housewives and old people able to do light housework only but able to go out shopping	D. Severe
V	Unable to undertake any paid employment Unable to continue any education Old people confined to home except for escorted outings and short walks and unable to do shopping Housewives able only to perform few simple tasks	
VI	Confined to chair or to wheelchair or able to move around in the house only with support from an assistant	
VII	Confined to bed	
VIII	Unconscious	

Source: Rosser, Kind and Williams (1982)

well-being descriptors might include: can the patient get up each morning, wash, dress, prepare a meal, clean the house, shop and go to work? Additionally, there may be measures of energy and pain/discomfort. Psychological well-being will include indicators of happiness and distress whilst the purpose of the social indicator is to measure social integration or loneliness.

An example of a set of descriptors from a quality of life measure is given in Table 2. These descriptors for disability and distress were valued by Rosser [3] to produce the matrix in Table 3. A state of no disability and distress is 1, dead is 0 and, as shown, the Rosser respondents viewed some health states as worse than dead.

Data overview

The volume and quality of input, process and outcome data are very poor. Most medical practices are unproven from a clinical and an economic perspective. Cochrane [4] argued that this was so twenty years ago. Sir Douglas Black [5] asserted that only 10% of medical practices are proven. Fuchs [6] has argued that 10% of health care expenditure worsens patient health, 10%

Table 3. The Rosser Valuation Matrix

Disability rating	Distress rating			
	A	B	C	D
I	1.000	0.995	0.990	0.967
II	0.990	0.986	0.973	0.932
III	0.980	0.972	0.956	0.912
IV	0.964	0.956	0.942	0.870
V	0.946	0.935	0.900	0.700
VI	0.875	0.845	0.680	0.000
VII	0.677	0.564	0.000	−1.486
VIII	−1.028	Not applicable		

Source: Kind, Rosser and Williams (1982)

has no effect and 80% improves health. Fuchs argues that the problem is that we do not know which therapies lie in the 10% and 80% categories!

Thus clinical and managerial decisions in health care systems are poorly informed by the products of systematic measurement. Choices are made on the basis of ignorance rather than science. "In God we trust, all others bring data." But how are these data to be used to improve choices about which patients will be diagnosed and treated, and which patients will be denied these actions even if they might benefit from them?

The economic approach to health care evaluation

Types of economic evaluation

Four types of economic evaluation are set out in Table 4: costing, cost benefit analysis, cost effectiveness analysis and cost utility analysis.

Costing involves the identification, measurement and valuation of what society gives up when a patient is diagnosed or treated. The costing is relative, i.e. two alternative procedures are costed and the outcome is assumed to be identical. The costing should be broad (i.e. identify the full costs to society) but often it is restricted to some subset of society, e.g. the hospital.

If the costing is restricted in this way, great care should be taken in its interpretation and use. A procedure that minimises costs for the hospital may shift patients and costs to primary care or households in a manner which does not facilitate the minimisation of the procedure's cost to society.

Cost benefit analysis (CBA) involves the identification, measurement and valuation of the costs and benefits of the alternatives, where both the cost and benefit streams are measured in monetary units (e.g. pounds or dollars). In the health care sector cost benefit analysis is often very difficult to carry

Table 4. Four types of economic evaluation

Type of Evaluation	How are costs measured?	How are outcomes measured?	How are outcomes valued?
Cost-mini-misation	Pounds (£)	Assumed equivalent	No valuation
Cost-effec-tiveness	Pounds (£)	Single variable common to the alternatives being evaluated but achieved to different degrees	Common units (e.g. years of life gained, days of disability avoided, units of blood pressure reduction)
Cost-benefit	Pounds (£)	Any effects produced by the alternatives	Pounds (£)
Cost-utility	Pounds (£)	Single or multiple effects, common or unique to the alternatives and achieved to differing degrees	Well years or QALYs

Source: Derived from Drummond et al. (1987)

out because many of the benefits are difficult to value in monetary terms, e.g. what is the value of pain reduction or blood pressure reduction?

As a consequence of these measurement problems with cost benefit analysis, a more restricted form of evaluation, cost effectiveness analysis (CEA), is carried out. The objective of a cost effectiveness analysis is to identify the costs of alternative routes to a common objective in diagnosis or treatment. Outcome is measured in a single common specific variable which can be achieved to varying levels e.g. additional life years achieved or millimetres of blood pressure reduction.

A problem with cost effectiveness analysis is that there is no common unit of outcome across evaluations and, as a consequence, it is difficult to compare the efficiency of a procedure which produces increased life years as opposed to a procedure which reduces blood pressure. The role of cost utility analysis (CUA) is to derive a common agreed unit of outcome which can be used to compare alternative interventions.

The outcome measure used in cost utility analysis combines the effects of enhancements in the length and quality of life to produce estimates of well years (WY) or quality adjusted life years (QALYs). The QALY effect of competing treatments can be derived by using 'expert judgements' about outcome in terms, for instance, of the Rosser matrix (Table 3) to produce Figure 2, i.e. experts are asked how patients move around the matrix pre and post treatment.

Treatment A produces benefits which decline in terms of quality of life over time, until the patient dies. Treatment B gives a different outcome. The difference between the two procedures, e.g. drug and surgical treat-

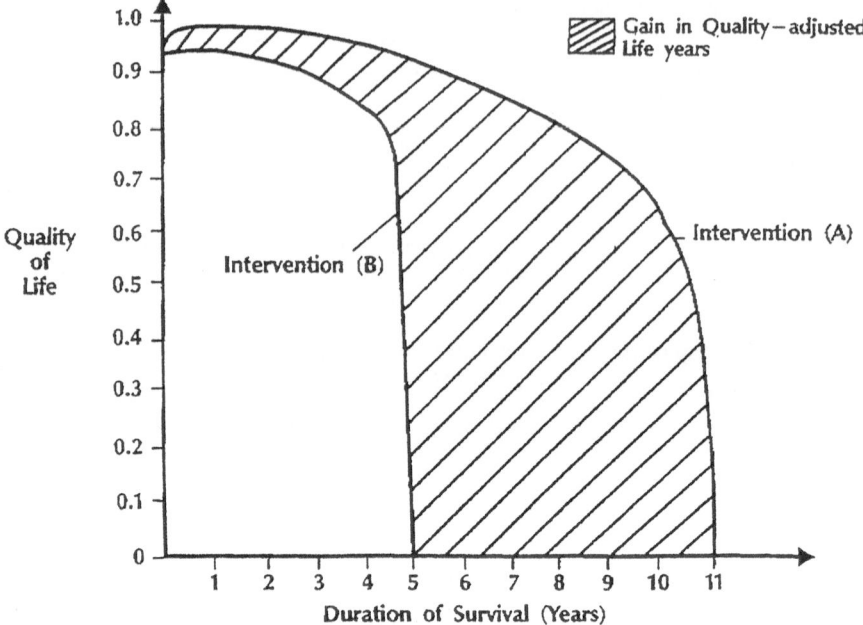

Fig. 2. The measurement of enhanced duration and quality of life

ment for angina, indicates the gain made from using one procedure rather than another and data about costs can then be used to determine the cost of producing incremental QALYs from different procedures. These gains (QALYs) can then be costed.

Some cost-QALY 'guestimates' using the Rosser quality of life measure are shown in Table 5.

The characteristics of a good economic evaluation

When judging the efficiency with which economic evaluations are completed it is necessary to scrutinise the following aspects of the study.
1. Is a sensible research question asked? It is foolish to address a general and poorly specified research question like "does endoscopy work?" Such a specification begs questions such as what are the characteristics of the endoscopic procedure (e.g. upper or lower gastrointestinal)? How is "work" measured?, and what is the comparitor, i.e. how does it work relative to what alternative?
2. Are the alternative interventions clearly identified? It is necessary in all studies to identify who does what to whom so that it is possible to judge whether the procedure is specific or able to be generalised.
3. Are all costs identified, measured and valued?
4. Are outcomes identified, measured and valued?

Table 5. The Cost per Quality Adjusted Life Year (QALY) of competing therapies: some tentative estimates

	Cost/QALY £ (Aug 1990)
Cholesterol testing and diet therapy only (all adults, aged 40–69) (Source A)	220
Neurosurgical intervention for head injury (B)	240
GP advice to stop smoking (C)	270
Neurosurgical intervention for subarachnoid haemorrhage (D)	490
Anti-hypertensive therapy to prevent stroke (ages 45–64) (C)	940
Pacemaker implantation (D)	1 100
Hip relacement (D)	1 180
Valve replacement for aortic stenosis (D)	1 410
Cholesterol testing and treatment (A)	1 480
CABG (IMVD, severe angina) (D)	2 090
Kidney transplant (D)	4 710
Breast cancer screening (E)	5 780
Heart transplantation (D)	7 840
Cholesterol testing and treatment (incrementally) of all adults 25–39 years (A)	14 150
Home haemodialysis (D)	17 260
CABG (1 vessel disease, moderate angina) (D)	18 830
CAPD (D)	19 870
Hospital haemodialysis	21 970
Erythropoeitin treatment for anaemia in dialysis patients (assuming a 10% reduction in mortality)	54 380
Neurosurgical intervention for malignant intracranial tumours (B)	107 780
Erythropoeitin treatment for anaemia in dialysis patients (assuming no increase in survival) (F)	126 290

5. Is the sample size of the study sufficient to demonstrate statistical significance (power) in the socio-economic variables? The variance of socio-economic data around the mean may be greater than for clinical data. As a consequence the sample size needed for socio-economic evaluation may be greater than that needed for clinical evaluation.

6. Are marginal values and outcomes identified? Information about the incremental effects of doing "one more" or "one less" of the procedure is more relevant for those seeking to determine whether to increase or decrease investment in a particular intervention area.

7. If costs and benefits accrue over time they have to be adjusted to take account of time preference e.g. if you are asked to pay £100 now or £100 in one year's time, you prefer the latter and the price that has to be paid to make you indifferent between paying now and in a year's time is the discount rate. Thus if you are indifferent to paying £100 now or £110 in a year's time, your discount rate is 10%.

It is usual to discount costs and benefits to take account of this time preference and most evalutions use discount rates of 5% or 6%. These values are somewhat arbitrary and there is some dispute about whether QALYs which arrive in different time periods should be discounted.

8. Sensitivity analysis of both the cost and outcome data are essential to identify the extent to which the results of the study change with different assumptions for estimating costs and outcomes.

Overview

There are all too few economic evalutions of diagnostic and therapeutic interventions. The methodology to carry out these studies is well established [7] and collections of case studies show how practice often falls short of principle [8].

The results of such studies are, however, becoming very important as information for the purchasers of health care who are seeking to make difficult choices between different interventions, all of which cannot be afforded in a cash conscious health care system. The purchasers are tending to target resources on those therapies which produce beneficial outcomes (QALYs) at low costs, i.e. those procedures towards the top of the listing in Table 5.

In North America similar crude approaches are being made to elicit priorities which will be used to ration health care resources. In Oregon, the Health Services Commission has prioritised 714 procedures and some examples are shown in Table 6. The legislative intends to apply funds to those procedures which it regards as efficacious, subject to incidence and budget constraints, i.e. procedures below 380 or so will not be financed with the result that people with full blown AIDS and neonates will not be treated. Such decisions are always difficult and the virtue of the economic approach it makes the making criteria explicit and subject to challenge. However, for an effective challenge to be made, better measurement is essential.

As resource management is developed, endoscopists in all countries will have to 'bid' for resources on the basis of cost and outcome data. A bid by a clinician to use a new procedure, for instance an antagonist, will have to be supported by good evidence about what costs will be saved and how the patient's quality of life will be enhanced. Rhetoric and noisy advocacy which is not sustained by good data are unlikely to facilitate the development of endoscopy in the 1990s.

Table 6. Orgeon priorities: some examples

1	Pneumoccal pneumonia
2	Tuberculosis
3	Peritonitis
23	Low birth weight (1250 g and over)
73	Low birth weight (1000–1249 g)
358	Low birth weight (500–749 g)
695	Liver cirrhosis
707	Terminal HIV, with 10% chance of 5 year survival
713	Low birth weight (less than 500 g, under 23 weeks gestation)

References

1. Priestman T et al (1989) The Royal College of Radiologists fractionation study. Clinical Oncology 1:63–66
2. McPherson K (1990) Why do variations occur? In: Anderson TF and Mooney G (eds.) The challenge of medical practice variations. Macmillan, London
3. Kind P, Rosser R, Williams A. (1982) Valuation of the quality of life: some psychometric evidence. In Jones-Lee MW (ed.) The value of life and safety, North Holland, Amsterdam
4. Cochrane AL (1972) Effectiveness and efficience. Nuffield Provincial Hospitals Trust, London
5. Black AD (1986) An anthology of false antitheses. Rock Carling 1984 Fellowship, Nuffield Provincial Hospitals Trust, London
6. Fuchs V (1984) The "rationing" of medical care. N Engl J Med 311 (24)
7. Drummond MF, Stoddart GL, Torrance GW (1987) Methods for the economic evaluation of health care programmes. Oxford University Press, Oxford
8. Drummond MF, Lowson K, Ludbrook A, Steele A (1985) Studies in economic appraisal in health care. Vol 2. Oxford University Press, Oxford

Further Reading

Department of Health, Standing Medical Advisory Committee (1990) Blood cholesterol testing: the cost-effectiveness of opportunistic cholesterol testing

Department of Health and Social Security (1986) Breast Cancer Screening, Forrest Report, HMSO, London

Hutton S et al (1990) The cost-effectiveness of the use of erythropoietin in the treatment of anaemia arising from chronic renal failure. Occasional Paper, Centre for Health Economics, University of York

Pickard J D et al (1990) Step towards cost-benefit analysis of regional neurosurgical care. Br Med J 301:629–635

Teeling Smith G (1990) The economics of hypertension and stroke. Am Heart J (119); 3 Part 2 (Suppl):725–728

Williams A (1985) Economics of coronary artery bypass grafting. Br Med J 249:326–329

CHAPTER 6

Report of Resource Management Workshop

Professor ALAN MAYNARD

The demand for medical interventions, diagnostic and therapeutic, exceeds their supply in many health care systems and therefore the "rationing" of resources is becoming more explicit. In some countries, e.g. France, Switzerland and Germany, these pressures are less obvious, although in these countries too resource managers are looking for data to contain cost inflation. The common problem, across all health care systems, public and private, is that data are few and choices are formulated in "data free" environments.

Costs of endoscopy

The main cost components in endoscopy are:
1. staff
2. drugs and consumables
3. equipment: instrument and light source
4. overheads (fixed and variable)

A questionnaire was designed to obtain information on procedures employed for both upper and lower gastrointestinal endoscopy, and to investigate their costs and quality of life (see page 76).

The questionnaire offered three options for which respondents attending the Workshop could supply data. Option A was endoscopy without sedation, option B was sedation with natural recovery, and option C was endoscopy with sedation and reversal with an antagonist. The questionnaire did not seek to elicit equipment costs which were assumed, arguably, to be similar across the world. The questionnaire was sent to eight endoscopists, and whilst all responded, some responses were incomplete.

The number of episodes per week for each procedure, option and respondent (1–8) are set out in Table 1 and varied considerably as did the duration of each procedure (Table 2).

The respondents found it very difficult to reply to all the questions, particularly about capital costs. The responses to the cost questions were poor and cost estimates were possible only for two respondents. These are reported in Tables 3 and 4.

Table 1. Number of endoscopic episodes each week on in-patients (IP) or out-patients (OP) with different treatment options (see text)

	Option A		Option B		Option C	
	IP	OP	IP	OP	IP	OP
Upper gastrointestinal endoscopy						
1.	0	0	5	14	0	0
2.	0	1	20	50	0	0
3.	0	2	18	28	2	0
4.	5	6	15	30	0	0
6.	13	23	0	0	0	0
7.	43	26	7	2	0	0
8.	30	18	11	0	0	2
Total	91	74	76	124	2	2
Colonoscopy						
1.	0	0	2	7	0	0
2.	0	0	10	25	2	1
3.	0	1	4	0	0	0
4.	0	0	10	20	0	0
6.	4	6	0	0	0	0
7.	6	2	20	10	0	0
8.	0	0	6	6	0	1
Total	10	9	52	68	2	2

Respondent No. 5 – does about 50 upper gastrointestinal endoscopies per week using option A only and about 10 colonoscopies per week using no sedation and no local analgesia

If we supplement responses with data from the British Society of Gastroenterology based on work carried out at the Freeman Hospital in Newcastle-upon-Tyne in the UK we can see cost ranges for both upper gastrointestinal endoscopy and colonoscopy, excluding instrument and overhead costs (Tables 5 and 6).

Following on from this we must attempt to include equipment costs and overhead costs, i.e. staff and buildings.

Using these data from the Freeman Hospital, we can see that an upper gastrointestinal diagnostic endoscopy will cost about £65 while a colonoscopy will cost £110 (Table 7).

What is absent from these data is the relationship between costs, volume and the quality of care in terms of the procedure and the outcome.

What is the relationship between average costs (Tables 5–7) and marginal costs? The marginal cost is the additional cost of performing the procedure on one or more patients and the results in terms of additional diagnoses. If an additional patient can be diagnosed without more staff and equipment in a given endoscopy session, then the marginal cost may be very low. If an

Table 2. Duration of endoscopic procedures (in minutes)

1. Upper gastrointestinal endoscopy

Procedure								Mean
Option A	10	–	15	10	4	5	3	7.8
Option B	15	5	20	10	–	5	5	10.0
Option C	15	–	20	–	–	–	5	13.3

Recovery

Option A	0	–	0	15	60	1	0	11.8
Option B	80	15	30	45	–	20	6	32.7
Option C	0	–	20	–	–	–	10	10

2. Colonoscopy

Procedure								Mean
Option A	–	–	20	22	20	20	20	20.4
Option B	30	15	35	22	–	20	30	23.7
Option C	30	–	35	–	–	–	30	30

Duration of Recovery Period

Option A	–	–	5	15	90	3.5	0	*
Option B	100	30	30	75	0	20	120	*
Option C	0	0	20	0	–	–	10	10

* There were large variations in options A and B for recovery from colonoscopy calculating means in this case would be of little value.

Note: Where range of times was reported, the mid-point of the range is used in the table.

additional patient can only be accommodated by the staffing and setting up of another session, then the cost may be very high.

What would happen if the unit was pressed into doing more endoscopies? How would the costs move? Many of the overhead costs might be fixed, and so they would not change. The staff might have to work a bit harder but there would be no great increases in staff costs. The equipment costs might not change very much.

The marginal costs may be very much lower than the average cost if the unit is not working at its fullest capacity and therefore more patients can be seen at a fraction of the cost of the Unit. However, if the volume of patients seen by the Unit is reduced then equipment, staff and overhead costs still have to be paid.

The natural analogy is an aircraft. If there is only an 85% load factor, adding another passenger is relatively cheap, since all that is needed is another allocation of in-flight catering. However, once 100% capacity is reached, the marginal costs of adding another passenger are very high – another plane will be needed.

Table 3. Cost of upper gastrointestinal endoscopy (£ sterling)

		Option A	Option B	Option C
Respondent 3				
Staff:	Medical	3.75	5.00	5.00
	Nursing	8.00	8.00	8.00
	Technical	0.00	0.00	0.00
	Administration	0.78	0.78	0.78
Drugs		0.23	2.07	16.29
Consumables		2.93	2.53	3.33
Total		£15.69	£18.38	£33.40
Respondent 8				
Staff:	Medical	2.00	1.33	1.33
	Nursing	4.50	36.00	5.00
	Technical	1.16	7.00	0.73
	Administration	0.58	0.88	0.88
Drugs		0.00	1.20	21.20
Consumables		0.12	0.20	0.20
Total		£8.36	£46.41	£29.34

These issues are critical to the management of endoscopy units because it is necessary to examine carefully the volume and the relationship between volume and costs. This is illustrated by a study which examined average and marginal costs in relation to the data testing for colonic cancer by analysis of occult blood in stools (guaiac) (Tables 8 and 9). What is the cost of identifying one more cancer of the colon taking into account both false positives and negatives? How far should this be taken, i.e. how many tests should be performed on the population in order to identify more cancers?

The yield fell off as more tests were done until there were hardly any cancers identified (Table 9). As more tests are carried out, the costs increase and rose from US $1175 to US $47 million. To identify one more cancer using the 6th test at a marginal cost of US $47 million does seem to be a relatively inefficient way of using resources.

If the same question had been asked of these data and the information computed in terms of average costs, they would have increased from US $1100 to US $2400 (Table 8).

Estimated costs must not only measure average costs but also marginal costs. That margin may be affected by another activity and that is whether the "input mix" can be altered, i.e. the combination of factors such as the labour and capital required to conduct the endoscopic intervention.

Table 4. Cost of colonoscopy (£ sterling)

		Option A	Option B	Option C
Respondent 3				
Staff:	Medical	7.50	9.00	9.00
	Nursing	4.50	12.00	12.00
	Technical	0.00	0.00	0.00
	Administration	0.78	0.78	0.78
Drugs		1.56	1.18	15.40
Consumables		4.80	4.80	4.80
Total		£19.14	£27.76	£41.98
Respondent 8				
Staff:	Medical	8.00	8.00	8.00
	Nursing	9.00	36.00	18.00
	Technical	1.75	7.00	3.50
	Administration	0.88	0.88	0.88
Drugs		0.50	2.50	32.50
Consumables		0.12	0.20	0.20
Total		£20.25	£54.58	£63.08

Table 5. Cost ranges for upper gastrointestinal endoscopy (£ sterling)

	Option A	Option B	Option C
Respondent 3 (Denmark)	16	18	33
Respondent 8 (Switzerland)	8	46	29
Freeman Hospital (UK)	NA	25*	NA

* It is assumed that the costs from the Freeman Hospital were based on the use of an agonist with natural recovery.

Table 6. Cost ranges for colonoscopy (£ sterling)

	Option A	Option B	Option C
Respondent 3	19	28	42
Respondent 8	20	55	63
Freeman Hospital	NA	38	NA

Table 7. Total costs for upper gastrointestinal diagnostic endoscopy (£ sterling)

Equipment cost	5.85
Staff/disposable	24.92
Overheads (fixed and variable)	34.02
Total	£64.79

Total costs for colonoscopy (£ sterling)

Equipment	14.04
Staff/disposable	38.25
Overheads (fixed and variable)	58.10
Total	£110.39

Table 8. Test results and costs for one to six stool guaiacs

No of Guaiacs	No of cancers found	No of cancers missed (false negatives)	False positives	Total cost of diagnosis (US $)	Average cost per cancer (US $)
1.	65.946	5.995	309	77,511	1,175
2.	71.442	0.4996	505	107,690	1,507
3.	71.990	0.0416	630	130,199	1,811
4.	71.938	0.0035	709	148,116	2,059
5.	71.941	0.0003	759	163,141	2,268
6.	71.942	0.00003	791	176,331	2,451

Table 9. Marginal results and costs for subsequent stool guaiacs

No of Guaiacs	Increase in numbers of cancers found	Increase in total cost (US $)	Marginal cost per cancer found (US $)
1.	65.946	77,511	1,175
2.	71.442	30,179	5,491
3.	71.990	22,509	49,146
4.	71.938	17,917	471,500
5.	71.941	15,025	4,038,978
6.	71.942	13,190	47,107,143

Quality assurance

What is the relationship between costs and quality in terms of process and outcome? There is considerable investment in "quality assurance". Quality assurance in some countries is being developed in a particularly ill thought out and unimpressive way with "top down" pronouncements which may be irrelevant to good patient care. The resource manager's demand for clinicians to raise volume (throughput) may leave little time to inform patients of everyday living immediately after the intervention. The value of these elements of the care process need to be evaluated so that process quality is not depressed by efforts to maximise volume.

Another aspect of quality is outcome. It was emphasised by the Workshop participants that a "normal finding" might be cost effective. An anxious patient whose worries are reduced by such an endoscopic procedure may use less consultation and drug time consequently. Thus the intervention mitigates concern, improves quality of life and saves health care resources.

The providers of diagnostic endoscopy have opinions about the quality of life effects of upper gastrointestinal and endoscopy procedures (Table 10). It is clear from these data that endoscopists are prepared to make guesses of the quality of life effects of different interventions, but it does seem that there was no consensus as to the quality of life implications of the procedures undertaken under the three predetermined options.

Table 10. Ranking for differences in quality of life

	Option A	Option B	Option C
Upper gastrointestinal endoscopy			
1.	2–3	1–2	1–2
2.	3	–	1
3.	1	2	2
4.	2	1	–
6.	1	3	–
7.	–	2	–
8*	–	–	–
Colonoscopy			
1.	2–3	1–2	1–2
2.	3	–	1
3.	2	1	1
4.	–	1	–
6.	1	3	–
7.	–	2	–
8*	–	–	–

* replied "No" to question on differences on quality of life.

The use of an antagonist

1. What recovery room and staff costs are freed up?
2. End of session use: is there relevant margin?

The high cost of the antagonist was clearly an obstacle to its wider use in the opinion of the Workshop participants. They agreed that its use would free up relevant and significant margins of recovery room capacity (overheads, fixed and variable) and staff time but there was agreement that these margins were limited and unquantified. Would the savings be real? Would capital be disposed of? If staff saving was redeployed in the Unit, then financial gains might be limited.

It was agreed that an increased use of an antagonist at the end of the session might be cost effective. This would move patients out of the Unit quickly so permitting staff to leave on time, thus avoiding overtime payments. With pressure to raise throughput in many units this margin might be quite significant and its importance needs careful investigation.

The implications of standard setting

If professional bodies i.e. the American Society for Gastrointestinal Endoscopy and the British Society of Gastroenterology set practice standards, their resource implications need to be taken into account. Standard setting without taking their costs into account is irresponsible. It will use resource and deprive potential patients of care from which they could benefit. Clearly risks and costs have to be balanced or traded off. It is too expensive to reduce risks to zero i.e. there is an efficient level of risk where doing a little bit more to reduce all risk is equal to the benefits. One Workshop member postulated that pulse oximetry, imposed by "risk-reducing" anaesthesiologists, is costly and not necessary if policies of low dose sedation and the use of supplemental oxygen were followed. Appropriate training and practice might obviate risks more cost effectively than mandating high technology procedures.

The data supplied by the eight respondents are illustrative rather than definitive and indicate the scope for future work.

a) Cost estimation is difficult in all health care systems. The eight "experts" from different health care systems who responded to the questionnaire found it difficult to obtain and manipulate cost data. Information about capital costs was particularly elusive.

b) Quality of life opinions show an expected variation but considerable agreement at this high level of generalisation.

Conclusions

Resource management at present is a crude and ubiquitous "game" whose consequences for medical practice and patient care are profound. It seems inevitable that in the 1990s we will see the development of
– cost "norms"
– "good practice" standards
– measures of quality

Where are the data to do this?

Ignorance will not inhibit resource managers who want to "guide" practice with explicit norms and targets. In the absence of data they will be guessed and guestimated similar to the events in Oregon.

The challenge to medical practitioners and the resource community is obvious. Better data about costs, processes and outcomes are needed. To meet this need, it is necessary to:
1. improve methodologies
    – what is a cost? (resource or finance)
    – how is quality of life to be measured?
2. apply existing methodologies in prospective trials.

Resource management can be based on ignorance or knowledge. There is a lot of the former and too little of the latter!

Resource Management Questionnaire

Introduction

i) Please complete this questionnaire in English
ii) Provide all financial data in pounds sterling (citing the exchange rate you used in your calculation).

1. Objective

Our objective is to identify the costs and outcomes (such as differences in the quality of life) of alternative sedative techniques for endoscopy. This questionnaire is limited to only diagnostic upper gastrointestinal endoscopy and diagnostic colonoscopy.

2. The alternative sedative techniques to be considered are:

Endoscopy with no sedation	Option A
Endoscopy with sedation with natural recovery	Option B
Endoscopy with sedation and reversal with antagonist	Option C

3. Explanation

In order to estimate the costs and effects of the three options it is necessary to acquire information on staffing, equipment, premises and their costs. Also it is necessary to acquire estimates of the relative effects of these technique options on the quality of life of patients. These data are often not easily accessible.

4. Information required

4.1 Availability of drugs

Are the following drugs available in your country and your hospital? Please delete as appropriate

	Country	Hospital
Midazolam	Yes/No	Yes/No
Diazepam	Yes/No	Yes/No
Flumazenil	Yes/No	Yes/No

4.2 Endoscopic episode

i) Activity

How many procedures are done by you/your team using each sedative technique in an average week according to how many are in-patients (IP) and how many are out-patients (OP)?

	Upper GI		Colonoscopy	
	IP	OP	IP	OP
Option A	. . . . .	. . . . .	. . . . .	. . . . .
Option B	. . . . .	. . . . .	. . . . .	. . . . .
Option C	. . . . .	. . . . .	. . . . .	. . . . .

ii) For out-patients only

For each technique, when does the patient receive information about diagnosis and treatment?
    Please delete as appropriate

	Option		
	A	B	C
Immediately after the procedure	Yes/No	Yes/No	Yes/No
Immediately before discharge	Yes/No	Yes/No	Yes/No
At a follow-up visit	Yes/No	Yes/No	Yes/No

iii) Indicate the average **duration of the endoscopic procedure** for each technique

Upper GI	Colonoscopy
Option A . . . . . mins	Option A . . . . . mins
Option B . . . . . mins	Option B . . . . . mins
Option C . . . . . mins	Option C . . . . . mins

iv) Indicate the average duration of the recovery period during which the patient needs observation for each technique

	Upper GI		Colonoscopy
Option A	..... mins	Option A	..... mins
Option B	..... mins	Option B	..... mins
Option C	..... mins	Option C	..... mins

4.3 Cost (exchange rate used £ =   )

i) Indicate the usual staff used in each endoscopy episode for each technique. The endoscopy episode starts with the endoscopy procedure (P) and includes the recovery period (R).

Upper GI

Option A

Staff – Medical

Grade	Number	Used in (P) or (R)	Time (mins)	Hourly wage rate
....
....
....
....

Staff – Nursing

Grade	Number	Used in (P) or (R)	Time (mins)	Hourly wage rate
....
....
....
....

Staff – Technical

Grade	Number	Used in (P) or (R)	Time (mins)	Hourly wage rate
....
....
....
....

Staff – Administrative

Grade	Number	Time (mins)	Hourly wage rate
. . . .			
. . . .			

Colonoscopy

Option A

Staff – Medical

Grade	Number	Used in (P) or (R)	Time (mins)	Hourly wage rate
. . . .				
. . . .				
. . . .				
. . . .				

Staff – Nursing

Grade	Number	Used in (P) or (R)	Time (mins)	Hourly wage rate
. . . .				
. . . .				
. . . .				
. . . .				

Staff – Technical

Grade	Number	Used in (P) or (R)	Time (mins)	Hourly wage rate
. . . .				
. . . .				
. . . .				
. . . .				

Staff – Administrative

Grade	Number	Time (mins)	Hourly wage rate
. . . .			
. . . .			

Upper GI

Option B

Staff – Medical

Grade	Number	Used in (P) or (R)	Time (mins)	Hourly wage rate
. . . .				
. . . .				
. . . .				
. . . .				

Staff – Nursing

Grade	Number	Used in (P) or (R)	Time (mins)	Hourly wage rate
. . . .				
. . . .				
. . . .				
. . . .				

Staff – Technical

Grade	Number	Used in (P) or (R)	Time (mins)	Hourly wage rate
. . . .				
. . . .				
. . . .				
. . . .				

Staff – Administrative

Grade	Number	Time (mins)	Hourly wage rate
. . . .			
. . . .			

Colonoscopy

Option B

Staff – Medical

Grade	Number	Used in (P) or (R)	Time (mins)	Hourly wage rate
. . . .				
. . . .				
. . . .				
. . . .				

Staff – Nursing

Grade	Number	Used in (P) or (R)	Time (mins)	Hourly wage rate
. . . .				
. . . .				
. . . .				
. . . .				

Staff – Technical

Grade	Number	Used in (P) or (R)	Time (mins)	Hourly wage rate
. . . .				
. . . .				
. . . .				
. . . .				

Staff – Administrative

Grade	Number	Time (mins)	Hourly wage rate
. . . .			
. . . .			

Upper GI

Option C

Staff – Medical

Grade	Number	Used in (P) or (R)	Time (mins)	Hourly wage rate
....				
....				
....				
....				

Staff – Nursing

Grade	Number	Used in (P) or (R)	Time (mins)	Hourly wage rate
....				
....				
....				
....				

Staff – Technical

Grade	Number	Used in (P) or (R)	Time (mins)	Hourly wage rate
....				
....				
....				
....				

Staff – Administrative

Grade	Number	Time (mins)	Hourly wage rate
....			
....			

Colonoscopy

Option C

Staff – Medical

Grade	Number	Used in (P) or (R)	Time (mins)	Hourly wage rate
. . . .
. . . .
. . . .
. . . .

Staff – Nursing

Grade	Number	Used in (P) or (R)	Time (mins)	Hourly wage rate
. . . .
. . . .
. . . .
. . . .

Staff – Technical

Grade	Number	Used in (P) or (R)	Time (mins)	Hourly wage rate
. . . .
. . . .
. . . .
. . . .

Staff – Administrative

Grade	Number	Time (mins)	Hourly wage rate
. . . .
. . . .

ii) Which drugs are used in each technique, e.g. local anaesthetics, sedatives, analgesics, antagonists? Please given generic and brand name of each drug.

Upper GI

Option A

Drug	quantity	unit cost
....		
....		
....		
....		
....		

Option B

Drug	quantity	unit cost
....		
....		
....		
....		
....		

Option C

Drug	quantity	unit cost
....		
....		
....		
....		
....		

Colonoscopy

Option A

Drug	quantity	unit cost
....		
....		
....		
....		
....		

Option B

Drug	quantity	unit cost
....
....
....
....
....
....

Option C

Drug	quantity	unit cost
....
....
....
....
....
....

iii) Which consumables are used in each option, e.g. needles, syringes, gloves, dressings etc?

Upper GI

Option A

Consumable	quantity	unit cost
....
....
....
....
....

Option B

Consumable	quantity	unit cost
....
....
....
....
....

Option C

Consumable	quantity	unit cost
....		
....		
....		
....		
....		
....		

Colonoscopy

Option A

Consumable	quantity	unit cost
....		
....		
....		
....		
....		

Option B

Consumable	quantity	unit cost
....		
....		
....		
....		
....		

Option C

Consumable	quantity	unit cost
....		
....		
....		
....		
....		
....		

4.4 Cost of recovery facilities

i)   How big is your recovery area (square metres)?

....

ii) How many beds?

....

iii) What is the capital/rental cost per year?

....

iv) What are the running/overhead costs (excluding staff)?

....

v)   How many hours is your recovery area used per week?

....

vi) What is the total number of patients using this area per week?

....

4.5 Is your choice and level of activity of technique influenced by any of the following considerations? Please delete as appropriate

	Option A	Option B	Option C
Medical staffing levels	Yes/No	Yes/No	Yes/No
Nursing staffing levels	Yes/No	Yes/No	Yes/No
Availability of equipment	Yes/No	Yes/No	Yes/No
Recovery facilities	Yes/No	Yes/No	Yes/No

4.6 Equipment

i) Are the following facilities or equipment available in your unit? Please delete as appropriate

Blood pressure monitor	Yes/No
ECG monitor	Yes/No
Pulse oximeter	Yes/No
Nasal oxygen	Yes/No

ii) Can you give the average cost of the following pieces of equipment in your country?

Non-invasive BP monitor ....

ECG monitor...

Pulse oximeter ....

iii) Which equipment do you use for each technique? Please delete as appropriate

	BP	ECG	Pulse oximeter	Nasal O_2
Option A	Yes/No	Yes/No	Yes/No	Yes/No
Option B	Yes/No	Yes/No	Yes/No	Yes/No
Option C	Yes/No	Yes/No	Yes/No	Yes/No

4.7 Outcomes

i) Is there a difference in mortality between the three techniques? YES/NO (Please delete as appropriate)
Is there a difference in morbidity between the three techniques? YES/NO (Please delete as appropriate)
If YES, please score each of the techniques
1 (best) to 3 (worst).

Upper GI

	Mortality	Morbidity
Option A	Score	Score . . . . .
Option B	Score	Score . . . . .
Option C	Score	Score . . . . .

Colonoscopy

	Mortality	Morbidity
Option A	Score	Score . . . . .
Option B	Score	Score . . . . .
Option C	Score	Score . . . . .

ii) Is there a difference in the quality of life between the techniques? YES/NO (Please delete as appropriate)
If YES, please score each of the techniques
Score 1 (best) to 3 (worst).

Upper GI		Colonoscopy	
Option A	Score	Option A	Score . . . . .
Option B	Score	Option B	Score . . . . .
Option C	Score	Option C	Score . . . . .

iii) Score the pain during the procedure for each technique
Score 0 (no pain) to 5 (severe pain)

Upper GI		Colonoscopy	
Option A	Score	Option A	Score . . . . .
Option B	Score	Option B	Score . . . . .
Option C	Score	Option C	Score . . . . .

iv) Score the discomfort during the procedure for each technique
   Score 0 (no discomfort) to 5 (severe discomfort)

	Upper GI		Colonoscopy
Option A	Score	Option A	Score .....
Option B	Score	Option B	Score .....
Option C	Score	Option C	Score .....

v) Score the nausea during the procedure for each technique
   Score 0 (no nausea) to 5 (severe nausea)

	Upper GI		Colonoscopy
Option A	Score	Option A	Score .....
Option B	Score	Option B	Score .....
Option C	Score	Option C	Score .....

vi) Score the patient's tension during the procedure for each technique
   Score 0 (no tension) to 5 (severe tension)

	Upper GI		Colonoscopy
Option A	Score	Option A	Score .....
Option B	Score	Option B	Score .....
Option C	Score	Option C	Score .....

vii) Score the discomfort during recovery for each technique
   Score 0 (no discomfort) to 5 (severe discomfort)

	Upper GI		Colonoscopy
Option A	Score	Option A	Score .....
Option B	Score	Option B	Score .....
Option C	Score	Option C	Score .....

viii) Score the anxiety during recovery for each technique
   Score 0 (no anxiety) to 5 (severe anxiety)

	Upper GI		Colonoscopy
Option A	Score	Option A	Score .....
Option B	Score	Option B	Score .....
Option C	Score	Option C	Score .....

ix) Score the mobility during recovery for each technique
   Score 0 (good mobility) to 5 (no mobility)

	Upper GI		Colonoscopy
Option A	Score	Option A	Score .....
Option B	Score	Option B	Score .....
Option C	Score	Option C	Score .....

Section 4: Medico-legal Aspects

Keynote Lecturer

Miss NICOLA DAVIES LLB
Barrister
3 Serjeants' Inn
London
United Kingdom

Member of British Society of Gastroenterology
Working Party on Informed Consent for Endoscopy

Workshop Director

Professor DAVID POSWILLO CBE DDS DSc MD FRCPath
Department of Oral and Maxillofacial Surgery
Guy's and St Thomas's Hospitals
London
United Kingdom

Chairman, UK Working Party on Anaesthesia, Sedation and Resuscitation
in Dentistry, 1990
Member of Board of Management, Medical Defence Union

Workshop Participants

Dr. D. COLIN-JONES, Portsmouth, United Kingdom
Dr. K. GAIL, Heppenheim, Germany
Dr. G. JUDMAIER, Innsbruck, Austria
Dr. J. P. MILLER, Manchester, United Kingdom
Dr. J. F. REY, St Laurent du Var, France
Professor Dr W RÖSCH, Frankfurt, Germany
Dr. J. C. SOUQUET, Lyon, France
Dr. W. M. WEINSTEIN, Los Angeles, California, USA

CHAPTER 7

Medico-legal Issues in Endoscopy

Miss Nicola Davies

Introduction

There are two main areas where endoscopists and their practice of endoscopy relate to the law, irrespective of whether the endoscopy is performed under sedation. Firstly, a doctor might be negligent in his treatment of the patient. Secondly, the endoscopist may fail to obtain informed consent.

Negligence

In England and Wales most legal actions arising from the professional conduct of a doctor in relation to his patient are brought in the civil courts and are founded in negligence.

In order to succeed in an action for medical negligence a patient must establish the following facts:
1. that a duty of care was owed by the doctor to the patient;
2. that the doctor was in breach of the appropriate standard of care imposed by the law;
3. that the breach of duty caused the patient harm or injury recognised by law as meriting compensation.

Duty of care

The duty arises from the fact that the doctor does something to a patient which is likely to cause physical damage unless it is done with proper care and skill. The duty is not reliant on the giving of a representation, undertaking or profession of skill. It exists whether or not there is a contract between the doctor and the patient.

If there is a contract between the doctor and the patient (private care) a term will be implied into the contract that the doctor will exercise reasonable skill and care in his treatment of the patient.

The scope of the doctor's duty will, in general terms, relate to all aspects of patient care including diagnosis [1], treatment [2] and advice [3].

Endoscopy is an invasive procedure usually performed when the patient is sedated. The patient's ability to take care of himself is compromised by reason of the sedation. The duty of the endoscopist is to ensure that the procedure is properly performed and that the patient is safe throughout the procedure.

This duty is not confined just to the performance of the endoscopy. If the endoscopist assumes the responsibility for sedating the patient or allows a second, more junior, member of the staff to carry out this task the endoscopist must ensure that the second person has sufficient training and experience not only to sedate the patient but also to deal with resuscitation should it be necessary. Similarly the endoscopist must, in these circumstances, be satisfied that the proper equipment for the endoscopy procedure, the administration of the sedation and adequate facilities for the resuscitation of the patient are readily available.

Standard of care

The question which is asked in medical negligence claims is whether the treatment fell below the standard of care reasonably to be expected of an ordinary competent doctor exercising the particular skill in question [4]. In order to answer this question the courts in England and Wales apply the following test: Did the doctor act in accordance with a practice accepted at the time as proper by a responsible body of medical opinion, even though other doctors adopt a different practice? Negligence is tested by the medical practice and knowledge prevalent *at the time* it is alleged to have occurred.

Medical witnesses

In order to make a decision on the primary question of negligence a judge will hear evidence from practitioners called on behalf of the patient and the doctor or health authority or hospital. In evaluating the evidence of the "experts" it is not for the judge to prefer one respectable body of professional opinion to another. If there is evidence given by an expert which is honestly held and truthfully given that the doctor acted in accordance with a responsible body of medical opinion that will be sufficient to defeat a claim in negligence.

Witnesses of fact

The treating doctors are, in the first instance, witnesses of fact. There are many cases which reach the courts some years after the alleged event and it will often be difficult for the doctor to remember the particular patient or some aspect of the treatment. In such a situation medical notes can be critical. The detailed note will not only serve as an *aide memoire* – it can positively reflect upon the general standard of care provided by a doctor. A sloppy or inadequate note will require an explanation.

Expert witnesses

Many doctors are now being asked to act as experts in cases. The role of the expert is two-fold: to educate the lawyers and to assist the judge. The lawyers look to the expert not only to explain what has occurred and why but to provide all relevant medical literature, e.g. textbooks and published papers, which demonstrate medical knowledge and practice at the time of the incident complained of. If guidelines have been produced by a professional society in the relevant field these should be provided as they can be used to illustrate prevailing practice and standards.

The impressive witness is the reasonable witness. However difficult the questions being asked, usually by the lawyer on the other side, the medical expert should not be pressured into expressing a view he has difficulty supporting. If a point is made in cross examination which the medical expert regards as reasonable, he should accept the point, it will add to this credibility as a witness.

One rule that is easy to state but less easy to follow is: when giving evidence, the question should be answered as it is asked.

The injury and damage suffered by the patient

Once again a judge will rely upon the medical experts to assist as to the nature and extent of the injury and the prognosis.

Consent

The issue of consent in the doctor/patient relationship can arise in two areas of the law:

Battery

Battery can be defined as the intentional application of force to the person of another, the force being harmful or offensive and being without the consent of that other and without lawful excuse. Perhaps the best description of the essence of harmfulness and offensiveness is that it lies in the unwanted nature of the touching [5].

In reality what this means is that a doctor acting in good faith with the entirely proper motive of seeking to help the patient is unlikely to be faced with a criminal or civil allegation of battery even though he may fail adequately to inform the patient of the nature and extent of the examination, investigation or procedure which he or she is to undergo.

Negligence

It is in this sphere that the issue of consent is most frequently litigated. The question usually raised being whether the patient was adequately informed of the nature, extent and risks of the procedure such as to enable that patient to give real consent.

The courts accept that the decision as to what risks should be disclosed is primarily a question of clinical judgment and require that the doctor acts in accordance with a practice accepted at the time as proper by a responsible body of medical opinion [3]. There is one exception to this rule, namely a risk of a substantial nature with grave adverse consequences. If such a risk exists and the patient was not so informed there is authority which would permit a judge to find as a fact the patient should have been told of the risk even if medical witnesses would not have so advised the patient [3].

Implied or express consent

Consent to treatment may be implied or express. Patients may imply agreement by compliant actions e.g. by offering an arm for a blood sample to be taken. Express consent is given when the patient confirms his agreement to a procedure in clear and explicit terms, verbally or in writing. Written consent should be obtained for any procedure or treatment carrying any substantial risk or side effect.

Endoscopy

Endoscopy is an invasive procedure and written consent should be obtained. A short written form which records the fact that the patient consents to undergo the procedure of endoscopy "the nature and purpose of which have been explained to me by Dr/Mr ..." will not be regarded by the courts as evidence that the patient has been fully and properly informed of the risks of the procedure.

When should the patient be informed of the nature and risks of the procedure?

The time to do this is not immediately before the endoscopic procedure. A patient has to be given a real opportunity to understand what is being said. Immediately before such a procedure a patient may be agitated, apprehensive or simply worried. The patient will not be in a state of mind which is conducive to listening to and easily understanding what is being said. The patient must have time to understand and appreciate the information and should be given a chance to ask questions or raise any doubts.

The patient, in-patient or out-patient, should be seen at least one day before the procedure and informed of the relevant facts. At this meeting it would be reasonable to ask the patient to sign the consent form. The patient should not be asked to sign the consent form immediately before the procedure because of his likely state of mind.

Who should inform the patient of the risks?

The responsibility for ensuring that the patient is adequately informed of the risks ultimately rests with the doctor carrying out the endoscopy. If this doctor delegates the task to another member of the medical staff the endoscopist must be satisfied that the second doctor has sufficient knowledge, experience and skill to do the job properly. The fact that the task has been delegated does not absolve the endoscopist from this responsibility.

Referred patients

Can a referral from another medical practitioner absolve the endoscopist from his duty to inform of the risks of the procedure?

Consultant referral

The patient is under the care of another consultant and that consultant has advised that a number of investigations should take place including that of endoscopy. The first consultant may have explained the risk of endoscopy as part of his general advice to and treatment of the patient. In these circumstances the endoscopist should ascertain that the risks have been adequately explained. It will not be necessary to repeat the exercise.

General practitioner referral

These frequently take place in open access clinics. The endoscopist cannot take for granted the fact that the general practitioner has explained the risks. An explanation of the risks must be given by the endoscopist or a doctor properly entrusted by him with this task.

Referral from another clinic in the same hospital

This patient should be treated in the same manner as the patient referred by a general practitioner.

Consent form

The forms used in many hospitals in England and Wales are general in their wording and do not specify the risks of the proposed procedure. Given the paucity of information contained in these forms it is in the doctor's interests to record in the medical records the risk of which he told the patient and any advice given.

Above all, the endoscopist should remember that as he or she is the person responsible for carrying out the invasive procedure, the safety of the patient is in his or her hands and that includes the right of the patient to know and hopefully understand what is involved.

Special cases

Children under the age of sixteen

In England and Wales a person of sound mind who has attained the age of sixteen years may give legally valid consent to surgical, medical or dental treatment or procedures (Section 8, Family Law Reform Act 1969). The Act does not state that a person below the age of sixteen cannot provide consent.

In the case of a child under sixteen the doctor must be satisfied that the child has sufficient understanding and intelligence to make the decision. The Department of Health advises doctors to make full note of the factors taken into account by the doctor in making his or her assessment of the child's capacity to give a valid consent [6].

Emergencies

In the case of a genuine emergency a doctor may proceed to do what is reasonably necessary to save life or prevent a deterioration in the patient in the absence of formal consent. The exception applies only to essential treatment.

Patients suffering from mental disorder

The presence of mental disorder does not itself imply incapacity nor does detention under the Mental Health Act. The capacity of a patient to give consent has to be judged in the light of the decision which has to be made and the mental state of the patient.

The Mental Health Act 1983 permits treatment of people detained in hospital in England and Wales under the powers of the Act without their consent when they are incapable of giving consent and the treatment is for mental disorder (Section 58, 63). The Act does *not* contain provisions enabling detained patients or patients suffering from a mental disorder to be given treatment for a physical disorder without consent. It is in this area that difficulties have arisen.

In 1989 the House of Lords considered the situation in relation to general medical and surgical treatment of people who lack the capacity to give consent [7]. In that case the Court held that save as to the treatment for their mental disorder the giving of medical treatment to mentally disordered persons was governed by common law and as such the Court had no jurisdiction to approve or disapprove the giving of this treatment. The lawfulness of the proposed operation depended on whether the treatment was in the best interests of a patient.

The Court provided some guidance in respect of such treatment:
i)   treatment which is necessary to preserve the life, health or well-being of the patient may lawfully be given without consent;
ii) the standard of care required of the doctor is that the doctor must act in accordance with a responsible body of medical opinion;

iii) in many cases it will not only be lawful for doctors on the ground of necessity to operate or provide medical treatment it will be their common law duty to do so;

iv) in the case of a patient suffering from a permanent or semi permanent state of mental disorder action properly taken may extend to routine medical and dental treatment.

References

1. Maynard v. W. Midlands RHA 1981 1 All ER 635 H.L.
2. Whitehouse v Jordan 1981 All ER 261 H.L.
3. Sidaway v Bethlem Royal Hospital Governors 1981 1 All ER 643 H.L.
4. Bolam v Friern Hospital Management Committee 1957 2 All ER 118
5. Kennedy and Grubb "Medical Law Text and Materials"
6. Department of Health HC (90) 22 Patient Consent to Examination or Treatment
7. Re F 1989 2 All ER 1025

CHAPTER 8

Report of Workshop on Medico-legal Issues

Professor DAVID POSWILLO

The medico-legal sword of Damocles hangs over all our heads and the eternal hope of every practitioner when learning of a medico-legal incident is "There but for the grace of God go I".

The endoscopist the sedationist and the law

In examining endoscopic practice internationally it seems anomalous that in France the endoscopist does not act as a sedationist. An anaesthesiologist must be present throughout the whole procedure, but that was not so for the other parts of the world. In Germany nurse sedationists are permitted and also in the United States just as they have nurse anaesthetists.

Negligence

Legal definition: "Actions in accordance with the practice as accepted as proper by a responsible body of persons skilled in that art is not negligent".

As long as the endoscopist acts in the same way as everybody who is equally skilled he cannot be found guilty of negligence. Since this is the law it cannot be avoided. It is important, when considering the risk-cost benefits of endoscopy, to take into account not only the cost of settling medical actions but also the cost of practitioners being anxious and concerned during the time course of the medico-legal process.

Procedural safety

A number of questions must be asked in the event of a collapse or a severe adverse reaction.
1. What went wrong and why?
2. Was the medical history known?
3. Was the sedation justified and was it necessary?
4. Was the agent or were the agents used correctly in this particular case?
5. Were trained persons present throughout the whole of the procedure and through the recovery?

6. Were there enough trained persons there and were they present during recovery?
7. Was resuscitation equipment available?
8. Was flumazenil readily available in the clinic?
9. Were trained and practised persons present who knew all about resuscitation?
10. Was each member of the team aware at once of all their duties with respect to managing a collapse or a severe adverse reaction?
11. Was there any failing on the part of anyone during or after the procedure which contributed to the collapse or severe adverse reaction?
12. Was there any failing on the part of anyone to resuscitate?

Intravenous access

Intravenous access should be available until the patient is fully recovered. This is not generally practised but it is very desirable and under no circumstances should cost inhibit the use of this particular technique.

Second appropriate person

It is the responsibility of the endoscopist to make certain that the people present are adequately trained and adequately maintained in their training. The law dictates that the hospital or the endoscopist should follow the practice considered standard in that locality. That is the yardstick by which the appropriate other person would be measured.

Informed consent and the law

A good way of overcoming many of the problems of obtaining informed consent is for each endoscopic unit to develop standard procedure forms which could be sent to the patient in advance of the endoscopy. The patient could read them, become aware of them, complete the standard procedure forms if they were required to answer any questions on medical history and return them to the endoscopist. The Workshop members believed that this information should be offered at least 24 hours before the procedure and furthermore the endoscopist in charge of the procedure must be prepared to answer any questions addressed to him in reasonable time before the procedure actually takes place. A standard procedure information leaflet or pamphlet combined with a consent form can be signed by the patient and returned to the endoscopist who can then, when he is quite satisfied that no questions need to be answered or all the questions asked have been answered, stamp it with an official stamp to say that all of the procedures required for informed consent have been carried out and completed. The Workshop strongly recommended this particular process for the procedure of gaining informed consent.

Information on likely adverse events

Patients should be informed of high incidence events and not necessarily of low incidence events. There are 'magic' figures of 10% for adverse events of a temporary nature and 0.5% events of a permanent nature which the medical profession thinks are the incidences above which the patient should be told. However, this is not the way the law views the situation. Every case is judged on the actions that occurred in that particular situation. A good rule to abide by when the endoscopist is asked "What shall I do doctor?" is to ask himself "What would I do if I were the patient or if a member of my family were the patient?". He tells the patient of all the alternative procedures including new techniques and doubtfully tested methods. If in the end the patient still asks "What would you do doctor?", the endoscopist resorts to 'gold standard' procedures. These are procedures which would be endorsed with absolute certainty by a reasonable number of colleagues should it ever come to the legal test.

To illustrate this point, what if a clinician has a patient with doubtful symptoms and he is worried about whether or not he should perform a cholecystectomy? In the presence of doubtful symptoms should he advise that no intervention takes place? The law occasionally says "Yes" and the clinician is on very solid ground in respect of the law if he has good justification for making that decision. What if the decision was wrong and there was an error in clinical judgement? The clinician is not going to be found negligent if he had reasonable grounds for his action according to a body of competent colleagues.

What should the clinician do about a new procedure? Should he immediately consider doing laparoscopic cholecystectomies because it is becoming fashionable? He should attempt a new procedure only if he feels, and can prove, that he is adequately trained. That is the standard by which he is going to be judged.

Consent for difficult problems

What should be done for the patient who says "Just go ahead and do whatever has to be done doctor, I do not want to know about it at all"? A record must be made in the notes, and witnessed by the patient that this is what he feels. The clinician is justified to go ahead and provide what he thinks is 'gold standard' treatment.

Mentally handicapped patients

Consent will have to be obtained from parents or guardians of mentally handicapped patients who are below the age of consent. On the other hand, the adult handicapped patient can usually make a decision him or herself but may need assistance from friends or relatives.

The battered patient

The battered or abused child is a different case altogether. A regulatory agency should never be informed about this matter unless the facts are 100% watertight.

What about the battered wife or husband? The legal view is that the adult is capable of making the decision for himself or herself. It is not for the clinician to make the decision on behalf of the battered person.

Mental illness

There are very considerable difficulties with obtaining consent from the mentally ill. Some of these patients may be in a position to make decisions for themselves but the clinician may have to go to the court for approval. If this were the standard procedure, the courts would be inundated with clinicians seeking permission to operate on patients who are mentally ill so therefore there must be an alternative. The law thinks a reasonable alternative is to ask a colleague to endorse the treatment plan, record it and witness it in the notes. This colleague should be on a different team so that there is no question of conflict of interest in that particular decision.

Consent obtained by trainees

What about the problem of consent being obtained by trainee doctors in the absence of their seniors? This is permissible provided that the senior clinician makes a checklist for the trainee who follows this checklist very carefully to make sure that all of the points have been made to the patient. Furthermore, the senior clinician should offer to answer any questions the patient may wish to address to him, in person or by telephone.

Consent obtained by practice nurses

In Germany consent by a practice nurse is legal, as is the nurse sedationist. The situation in other European countries may change after 1992.

Conclusion

Lord Denning, a very controversial and extremely articulate High Court judge in the United Kingdom said that all people face risks. We all face a risk when we get into a taxi. We are not sure that the taxi is not going to be involved in an accident, that the motor is going to perform effectively or that the brakes are in good working order. The taxi driver does not ask your informed consent before we get in. He takes us to the station or the airport so we take risks – everybody takes risks. Doctors and patients must take risks, and the degree of risk must continue as long as medical advances continue. If there is no risk there will never be any advance.

Subject Index